Remote Warfare

Interdisciplinary Perspectives

EDITED BY

ALASDAIR MCKAY, ABIGAIL WATSON AND
MEGAN KARLSHØJ-PEDERSEN

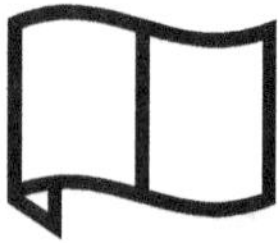

E-INTERNATIONAL RELATIONS PUBLISHING

E-International Relations
www.E-IR.info
Bristol, England
2021

ISBN 978-1-910814-56-7

Production: Michael Tang
Cover Image: Ruslan Shugushev/Shutterstock

A catalogue record for this book is available from the British Library.

E-IR Edited Collections

Series Editor: Stephen McGlinchey
Books Editor: Bill Kakenmaster
Editorial assistance: Amira Higazy, Bárbara Campoz Diniz, Franny Klatt, Muskan Yadav.

E-IR's Edited Collections are open access scholarly books presented in a format that preferences brevity and accessibility while retaining academic conventions. Each book is available in print and digital versions, and is published under a Creative Commons license. As E-International Relations is committed to open access in the fullest sense, free electronic versions of all of our books, including this one, are available on our website.

Find out more at: http://www.e-ir.info/publications

About E-International Relations

E-International Relations (E-IR) is the world's leading open access website for students and scholars of international politics, reaching over three million readers per year. In addition to our books, our daily publications feature expert articles, reviews and interviews – as well as student learning resources. The website is run by a non-profit organisation based in Bristol, England and staffed by an all-volunteer team of students and scholars.

http://www.e-ir.info

Abstract

Modern warfare is becoming increasingly defined by distance. Today, many Western and non-Western states have shied away from deploying large numbers of their own troops to battlefields. Instead, they have limited themselves to supporting the frontline fighting of local and regional actors against non-state armed forces through the provision of intelligence, training, equipment and airpower. This is remote warfare, the dominant method of military engagement now employed by many states. Despite the increasing prevalence of this distinct form of military engagement, it remains an understudied subject and considerable gaps exist in the academic understanding of it. Bringing together writers from various backgrounds, this edited volume offers a critical enquiry into the use of remote warfare.

Acknowledgments

The editors would like to thank the many people who gave up their time and helped with the review process for this book. Some of them have preferred to remain anonymous and are not named here. None of them have responsibility for any of the opinions (or errors) in this book, which are the authors' own: Feargal Cochrane, Florence Gaub, Hijab Shah, Ian Davis, James DeShaw Rae, James Igoe Walsh, James Strong, John Hutchinson, Kjetil Enstad, Louise Wiuff Moe, Marco Jowell, Maria Ryan, Nigel White, Nikolaos van Dam, Philippe Frowd, Ryan Jenkins and Simone Papale.

Alasdair McKay is Senior Editor at Oxford Research Group. He holds undergraduate and postgraduate degrees from the universities of Manchester and Aberystwyth. He has edited several books for E-International Relations, including *Nations under God: The Geopolitics of Faith in the Twenty-First Century* (2015) and *Into the Eleventh Hour: R2P, Syria and Humanitarianism in Crisis* (2014).

Abigail Watson was ORG's Research Manager and is now a Conflict and Security Coordinator at Saferworld. She researches and presents on the military, legal and political implications of remote warfare. She has produced over 50 blogs, briefings, podcasts and reports on this subject and has presented her work at institutions like Chatham House, the Cabinet Office, the Stabilisation Unit, the Ministry of Defence and the International Committee of the Red Cross. Her work has been featured in *INews, The Times, the History Channel* and *Just Security*.

Megan Karlshøj-Pedersen is a Research and Policy Officer at the ORG. She researches the shift towards remote warfare, advocating for more transparent and accountable approaches. She has examined the challenges of contemporary remote warfare in co-authored hard-hitting reports, including on the challenges of the UK's engagements in the Sahel and Horn of Africa and the strategic importance of protecting civilians in contemporary conflict. Her blogs, briefings, and articles have been published in outlets such as the *Small Wars Journal, Defense One, openDemocracy, International Review* and others. She has also presented on these topics at many conferences and workshops in the UK and internationally.

Contents

Contributors

Baraa Shiban is a Middle East and North Africa Caseworker at Reprieve. He also works as a political consultant to the Yemeni Embassy in London. Prior to this role, Baraa was Reprieve's Yemen Project Coordinator, investigating drone strikes across Yemen. Baraa also served as a youth representative in Yemen's National Dialogue, a convention tasked to negotiate Yemen's new constitution and reviewing its laws. Baraa is a Business Administration graduate and was involved with a number of civil society organisations in Yemen from 2006–2011. In 2011, he participated in peaceful demonstrations against Ali Abdullah Saleh, helping run a media centre in Sana'a's Change Square. As Reprieve's Yemen Project Coordinator, Baraa interviewed witnesses and civilian victims of US air strikes around Yemen, including people from Rada'a, Khashamir, Wessab, and towns in Ayban and Marib. Baraa speaks Arabic and English.

Camilla Molyneux is a research consultant and Policy Advisor at the All-Party Parliamentary Group on Drones, with a focus on sustainable security, civilian protection in armed conflict and parliamentary scrutiny over the use of force. She is also an independent analyst and researcher specialising in remote warfare and its impact on civilian populations and Yemen. Camilla has previously worked on defence, foreign policy and human rights for a civil society organisation and as the Human Rights Officer at the Norwegian Embassy to Saudi Arabia, also covering Yemen, Bahrain and Oman.

Christopher Kinsey is a Reader in Business and International Security with King's College London, Defence Studies Department at the Joint Services Command and Staff College, where he teaches military officers from around the world. His research examines the role of the market in conflict. Dr Kinsey has published widely on the subject through books, book chapters and articles in leading academic journals. He has also presented papers to the UN Intergovernmental Working Group on PMSCs (private military and security contractors), NATO and the EU Sub-Committee on Human Rights. Dr Kinsey's present work looks at the regulation of private security companies, the impact of contracted logistical support to military expeditionary operations, and mercenary operations in Africa during the Cold War. His previous books include *Corporate Soldiers and International Security* (London: Routledge, 2006); *Private Contractors and the Reconstruction of Iraq: Transforming Military Logistics* (London: Routledge, 2009); and the edited volumes, *Contractors and War: The Transformation of United States' Military and Stabilization Operations* (USA: Stanford University Press), *The Routledge Research Companion to Security Outsourcing* (London: Routledge, 2016), and *Embassies Under Siege: Diplomatic Security Policies Compared* (California: Stanford University Press, 2019).

Daniel Mahanty is the Director of the US Program for the Center for Civilians in Conflict (CIVIC). Through research and advocacy, he promotes the adoption of US Government policies and practices that serve to limit the harm experienced by civilians in armed conflict. Prior to CIVIC, he spent 16 years in the US Department of State, where he created and led the Office of Security and Human Rights. In this role, he oversaw the integration of human rights in US security assistance and arms sales policies and coordinated US Government efforts to prevent the recruitment and use of child soldiers. Mahanty holds a master's degree in national security policy from Georgetown and a bachelor's degree in economics from George Mason University. He is an adjunct professor at Kansas University Center for Global and International Studies at Ft. Leavenworth, Kansas and a non-resident Senior Associate at the Center for Strategic and International Studies.

Delina Goxho is an independent security analyst working for the Open Society Foundations, currently covering the Sahel, security sector assistance and armed drones. Prior to this, Delina was an analyst and researcher for Open Society's portfolio on armed drones, the NATO Parliamentary Assembly Defence and Security Committee and the Task Force Iran at the European External Action Service.

Helene Olsen is a doctoral candidate in the Department of War Studies at King's College London. She teaches various courses in both the Department of War Studies at King's and at the Joint Services Command and Staff College in Shrivenham. Currently, her research focuses on the use of mercenaries and their perceived illegitimacy, as well as the privatisation of security and military functions more broadly. Her work features in *Violent Nonstate Actors in Conflict* (Howgate Publishing, 2021).

Hendrik Huelss is Assistant Professor at the Centre for War Studies, Department of Political Science and Public Management, University of Southern Denmark. He is a researcher in the ERC-project AUTONORMS, investigating the implications of emerging autonomous weapons systems for international relations' (IR) norms. His primary research interests include governmentality studies in IR, the EU's external relations, the role of norms and technologies in IR. Hendrik's work has been published in journals such as *Review of International Studies, International Theory, International Political Sociology* and *Journal of International Relations and Development.*

Ingvild Bode is Associate Professor at the Centre for War Studies, University of Southern Denmark. She is also Principal Investigator for the ERC-funded project AUTONORMS, where she explores how weaponised artificial intelligence will change international norms governing the use of force. Ingvild is the author of *Individual Agency and Policy Change at the United Nations* (Routledge, 2015) and the co-author of *Governing the Use-of-Force: The Post-9/11 US Challenge on International Law* (Palgrave, 2014, with Aiden Warren). She has published in journals such as the *European Journal of International Relations, Global Governance, International Studies Perspectives*, and *Contemporary Security Policy.*

Jennifer Gibson is Head of Assassinations Project, Reprieve. She works closely with civilian victims of covert US drone strikes, investigating their cases and trying to help them get accountability, whether through the courts of law or the courts of public opinion. She has testified about her work before both the British Parliament, the European Parliament and the US Congress.

Jolle Demmers is Full Professor in Conflict Studies, co-founder of the Centre for Conflict Studies and the Director of the History of International Relations section of Utrecht University. She is the author of *Theories of Violent Conflict* (Routledge, second edition, 2017). Together with Lauren Gould, she is the founder of The Intimacies of Remote Warfare programme, among their recent publications is 'An Assemblage Approach to Liquid Warfare' (Security Dialogue 2018).

Joseph Chapa is an officer in the US Air Force and holds a PhD (DPhil) in philosophy from the University of Oxford. His areas of expertise include just war theory, military ethics, and especially the ethics of remote and autonomous weapons. His doctoral research investigates an individual rights-based account of just war theory. He is a senior pilot with more than 1,400 pilot and instructor pilot hours, many of which were flown in support of major US combat and humanitarian operations.

Julian Richards has spent nearly twenty years working for the British Government in intelligence and security policy, before co-founding the Centre for Security and Intelligence Studies at the University of Buckingham where he is currently the Director. His research interests include intelligence machinery and governance; and counter-terrorism policy in a range of regional and global contexts.

Lauren Gould is Assistant Professor in Conflict Studies at the Centre for Conflict Studies at the History of International Relations section of Utrecht University. Together with Jolle Demmers, she is the founder and the project leader of The Intimacies of Remote Warfare programme. The programme aims to inform an academic as well as public debate on the intimate realities of the remote wars waged in our name.

Malte Riemann is Senior Lecturer in the Department of Defence and International Affairs at the Royal Military Academy Sandhurst and a Visiting Research Fellow at the University of Reading. His research focusses on the historicity of violent non-state actors, practices of militarisation, remote warfare, and the relationship between public health and conflict. Together with Norma Rossi, he is co-founder and series editor of the *Sandhurst Trends in International Conflict* (Howgate Publishing). His work has been published in various journals, including *Journal for Global Security Studies*, *Critical Public Health*, *Small Wars Journal*, *Peace Review*, and *Discover Society*. He most recently published a monograph in German on the transformation of war titled *Der Krieg im 20. und 21. Jahrhundert* (Kohlhammer Verlag 2020).

Norma Rossi is a Senior Lecturer in the Department of Defence and International Affairs at the Royal Military Academy Sandhurst and a Visiting Research Fellow at the University of Reading. Her research focuses on the co-production of authority and subjectivity at the intersection between legal, political, and social dimensions with a specific focus on war and security. She has published on the political character of organised crime and its relation to state building, the impact of counter-organised crime measures on states' practices and discourses on security, the changing character of war, and the value of further education in conflict resolution. Together with Malte Riemann, she is co-founder and series editor of the *Sandhurst Trends in International Conflict Series* (Howgate Publishing). Norma has published in various journals including *Global Crime*, *Small Wars Journal*, *Journal of Civil Wars*, *E-International Relations*, and *Peace Review* (forthcoming). She most recently published a chapter in *Law, Security and the Perpetual State of Emergency* (Palgrave MacMillan 2020).

Rubrick Biegon is a Lecturer in International Relations in the School of Politics and International Relations at the University of Kent. His current research interests include US foreign policy, international security, and remote warfare. His recent articles have appeared in *Global Affairs*, *The Chinese Journal of International Politics*, and *International Relations*, among other outlets.

Sinan Hatahet is a Senior Associate Fellow at Al Sharq Forum. He is currently a consultant working with a number of think tanks on Syria. His research is concentrated on governance and local councils, anti-radicalisation, Islamism, the Kurdish National Movement, and the new regional order in the Middle East. He previously worked as the Executive Director of the Syrian National Coalition media office from its establishment in late 2012 until September 2014.

Tom Watts is a Teaching Fellow in War and Security at Royal Holloway, University of London with research specialisations in American foreign policy, military assistance programs, and lethal autonomous weapons systems. His PhD thesis asked what the Obama administration's military response against al-Qaeda's regional affiliates in the Arabian Peninsula, the Horn of Africa and the Sahel tells us about the means and goals of contemporary US military intervention in the Global South. Working within the historical materialist tradition, it advances a more critical reading of these processes which places military assistance programs and the reproduction of 'closed frontiers and open-doors' at the centre of its analysis.

Introduction

ALASDAIR MCKAY

Modern warfare is becoming increasingly defined by distance. Today, instead of deploying large numbers of boots on the ground, many Western and non-Western states have limited themselves to supporting the frontline fighting of local and regional actors. To counter non-state armed groups like Boko Haram, al-Shabaab and Islamic State, states have engaged mostly through the provision of intelligence, training, equipment, airpower and small deployments of special forces. This is remote warfare, the dominant method of military engagement employed by states in the twenty-first century.

However, despite the increasing prevalence of this distinct form of military engagement across Africa, Asia, the Middle East and parts of Europe, it remains understudied as a topic and considerable gaps exist in the academic understanding of it. This, in part, explains why it is also a subject clouded by several dangerous myths. There are assumptions and common narratives in certain political and military spheres that remote warfare is politically risk-free and does not produce significant civilian harm (see Knowles and Watson 2017, 20–28; Walpole and Karlshøj-Pedersen 2019, 2020). This edited volume seeks to start filling those gaps, challenging the dominant narratives and subjecting remote warfare to greater scrutiny.

The chapters in this volume come from papers presented at an academic conference entitled *Conceptualising Remote Warfare: The Past, Present and Future* which was hosted at the University of Kent in April 2019.[1] Co-organised by the Oxford Research Group, an international security think-tank, and the School of Politics and International Relations at the University of Kent, the event brought together a diverse range of participants from various academic disciplines and professional backgrounds, including the military, civil society and non-governmental organisations.

The event was organised to help foster a more holistic understanding of this

---

[1]   Podcasts of the event panel discussions can be listened to here https://www.oxfordresearchgroup.org.uk/pages/category/event-podcast-conceptualising

trend in military intervention; to promote greater dialogue between different research and practitioner communities working in remote warfare; and to encourage reflection on the recent debates surrounding remote warfare. Following three days of presentations and panel discussions, the key takeaways from the symposium were that the intellectual and professional pluralism of 'remote warfare scholarship' is one of its strongest attributes and that the inclusion of diverse viewpoints in debates surrounding its use will be crucial to understanding this phenomenon going forward (Watts and Biegon 2019).

Showcasing some of the conference's intellectual diversity, this book includes contributors from various backgrounds and disciplines who critically engage with the key debates and themes surrounding the use of remote warfare. There are fourteen chapters in the book, which are bound together by three interlocking themes.

The first theme is the opacity surrounding the practices associated with remote warfare and its implications for democratic states. Remote warfare essentially allows military operations to be conducted mostly away from mainstream public and political discussion. But this has serious consequences because it can undermine democratic controls designed to hold to account decisions to use force abroad. This also has ramifications in the theatres where remote warfare is conducted. Several of the chapters attend to the lack of transparency and accountability surrounding the use of remote warfare.

The second, and related, theme of the book concerns the long-term implications of remote warfare for peace and stability in states where it is used. Several chapters in the volume demonstrate how, despite occasionally yielding some forms of short-term tactical success, remote warfare can be detrimental to long-term peace and stability in several places where it is employed.

This dovetails into the third theme – the relationship between remote warfare and civilian harm. While remote military interventions are often portrayed as 'precise' and 'surgical', the various facets of remote warfare can, and often do, lead to civilian casualties. Empowering local partners – who may not have the capacity or sufficient interest in implementing strong protection of civilian mechanisms – and relying on airpower creates significant risks for local populations. Several of the chapters question the belief that it is possible to do remote warfare 'cleanly.' They highlight the various consequences that utilising remote warfare can have on civilian populations.

The civilian harm issue in remote warfare is also closely connected to debates about how the growth of technology impacts warfare. The final chapters look at how the dawn of future technologies such as artificial intelligence may shape remote warfare in the years ahead.

Although the book is designed to bring together a wide range of views, it should not be thought of as the 'encyclopaedia of remote warfare', covering every case of this phenomenon across time and space. After all, there is only so much that can be done in a text of this kind and there are always interesting angles and case studies left unexplored in every book. Rather, the long-term goal of this book is to create a text that students and scholars can learn from, critically engage with and potentially build upon in future work.

Book structure and chapter summaries

The opening chapter, written by the book's editors, is intended to serve as a primer on remote warfare and provide some conceptual clarity on the subject. It sketches out an overview of the concept, the broader debates surrounding its use and the problems that this type of engagement can yield. It serves as both an introduction for readers unfamiliar with some of the thematic areas and as a critical analysis.

The chapter by Jolle Demmers and Lauren Gould continues the discussion. It fleshes out the reasons why several liberal democracies have turned to remote warfare as an approach. The authors posit three key reasons for the turn to remote approaches to intervention: democratic risk aversion, technological advancements and the networked character of modern warfare. It then outlines some of the major consequences of this shift. The chapter explores how the secrecy surrounding remote warfare, and the way its practiced, attempts to remove war from public debate and potentially makes states more violent rather than less violent.

The opaque world of intelligence sharing, and the dilemmas the practice yields for states in the contemporary security environment, serves as the focus of Julian Richards' article. Using the UK in the post-9/11 environment as a case study, the chapter looks at the benefits and pitfalls of intelligence sharing in the current era. It considers how far the benefits to be gained for states with international intelligence sharing relationships outweigh the risks to democratic states and societies.

Transparency and accountability, and their importance in democratic states' use of force, feature heavily in Christopher Kinsey and Helene Olsen's chapter on the role of military and security contractors in remote warfare. The article provides an overview of the use of security contractors by states in the contemporary international security environment, the rationales for employing them and the potential problems in doing so. It suggests new ways for states to move forwards when using contractors in the future. The authors highlight why a more open debate surrounding the use of contractors will be essential in the future.

Norma Rossi and Malte Riemann's chapter also looks at the use of private contractors by states and examines the social and political consequences of this for the countries employing them. The article explores how the use of contractors, and remote warfare more broadly, by states has reshaped modes of remembrance, duty and sacrifice in societies. This has made war appear more distant and less visible within democratic societies.

As noted earlier, the lack of transparency surrounding remote warfare can also have significant impacts on the societies where remote warfare is being conducted. Delina Goxho's chapter focuses on a prominent arena of remote warfare, the Sahel. Goxho explores the interventionist footprints in the region, including the use of remote warfare tactics by the US and France. The article illustrates that the clandestine nature of remote warfare used by intervening states places a strain on local communities in the Sahel, who are ill-informed of military operations in the region. This is having a negative impact on peace and stability in the region. Offering some hope for the future, the chapter explores how the European Union could serve as a peace broker in the Sahel.

The realities of remote warfare for civilian communities on the ground are exposed further in Baraa Shiban and Camilla Molyneux's chapter. The chapter focuses on Yemen, where a civil war and several forms of intervention are ongoing. Drawing on their fieldwork in the country, which involved interviewing local populations, they illustrate the harm generated by remote warfare operations, in the form of US Special Forces raids and drone strikes, in the country.

The crisis in Yemen and the broader issue of civilian harm in remote warfare also serves as the prompt for Daniel Mahanty's chapter. Mahanty examines the dangers for civilian populations that can stem from relying on strategic partners. The chapter then sets out a summary of a framework, developed by the Center for Civilians in Conflict, for how militaries such as the US might assess the potential for human rights violations and civilian casualties when undertaking security cooperation activities with partners.

Moving the focus to the Horn of Africa, Rubrick Biegon and Tom Watts' chapter continues the discussion on security cooperation by examining US capacity building activities in the region. By examining the use of 'advise, train and assist' missions by the US in parts of the African continent, with a particular focus on Somalia, the chapter unearths how security co-operation seeks to fulfil both security and political goals. Focusing on this aspect of security cooperation, the authors believe, can help us better understand why security cooperation is still undertaken by the US in the continent despite the high-profile military failures it has caused.

For the past decade, the conflict in Syria has been a prominent arena of remote warfare, where several international actors have supported local forces and militias. Sinan Hatahet's contribution discusses the effects that the use of remote warfare has had on the state and the region. As Hatahet shows, the use of these activities by intervening forces has contributed heavily to creating distrust between domestic and international actors and destabilising the region with disastrous consequences for civilian populations.

While supporting local forces may well be a weapon of choice for states moving forward, drones will undoubtedly play an important role in states' overseas activities. Their use will continue to present all manner of political, legal and moral questions. Jennifer Gibson's paper examines the case of Faisal bin ali Jaber, a Yemeni engineer whose innocent brother-in-law and nephew were killed in a US drone strike. This strike was undertaken using a targeting algorithm which uses metadata to help decide who is a target. The chapter addresses some of the troubling questions that arise as big data and remote warfare converge. It then examines how using algorithms to make life and death targeting decisions relates to international humanitarian law, particularly the protection of civilians in conflict.

As the technology evolves, broader questions arise around how this may impact human control over the use of force. Joseph Chapa's chapter analyses the relationship between armed drones and human judgment, specifically as it pertains to targeting decisions. Drawing upon interviews with armed drone pilots, Chapa argues that, though the physical distance between aircrew and the targets on the ground presents difficult challenges, pilots can still apply human judgment when undertaking strikes. The chapter also warns how the advent of new technologies such as artificial intelligence could challenge this capacity for human judgement in remote warfare.

Following the chapters by Chapa and Gibson, Ingvild Bode and Hendrik Huelss' contribution investigates the potentially game-changing role of artificial intelligence in autonomous weapons systems (AWS). The authors argue that the development of weapon systems with greater autonomy may challenge the existing norms governing the use of force. This could have highly problematic political and ethical consequences.

To bring these issues together, the book's conclusion opens up a further discussion of what the future of remote warfare might look like. Here, the chapter explores how the rise of great power competition may influence the use of remote warfare and then factors in how the recent COVID-19 pandemic may impact matters. The conclusion then points to several directions that researchers could explore in future studies of remote warfare.

References

Knowles, Emily, and Abigail Watson. 2018. *Remote Warfare: Lessons Learned from Contemporary Theatres*. Remote Warfare Programme, Oxford Research Group.

Oxford Research Group. 2019. 'Conceptualising Remote Warfare'. Event Podcasts. https://www.oxfordresearchgroup.org.uk/pages/category/event-podcast-conceptualising

Walpole, Liam, and Megan Karlshøj-Pedersen. 2019. 'Remote Warfare and the Practical Challenges for the Protection of Civilians Strategy.' Oxford Research Group.

———. 2020. *Forging a New Path: Prioritising the Protection of Civilians in the UK's Response to Conflict*. Oxford Research Group. https://www.oxfordresearchgroup.org.uk/forging-a-new-path-prioritising-the-protection-of-civilians-in-the-uks-response-to-conflict

Watts, Tom, and Rubrick Biegon. 2018. 'Conceptualising Remote Warfare: Past, Present and Future.' Oxford Research Group.

1

Remote Warfare: A Critical Introduction

ABIGAIL WATSON AND ALASDAIR MCKAY

In the twenty-first century, remote warfare has been the most common form of military engagement used by states. But as the introduction noted, it remains a poorly understood concept. To some readers it may even be an unfamiliar term. To set the scene for the later contributions, this opening chapter acts as a critical conceptual primer on remote warfare. The chapter first outlines the key techniques involved in remote warfare. It then provides a glimpse of what it looks like in practice, where it is being used and by whom. After this, the chapter proceeds to examine remote warfare's relationship with the changing character of the war debate. Drawing upon research by Oxford Research Group (ORG), the penultimate section critically engages with some of the key challenges with its use. The chapter then offers some concluding remarks.

What is remote warfare and what does it consist of?

As the name hints, remote warfare refers to an approach used by states to counter threats at a distance. Rather than deploying large numbers of their own troops, countries use a variety of tactics to support local partners who do the bulk of frontline fighting. In this sense, the 'remoteness' comes from a country's military being one step removed from the frontline fighting (Knowles and Watson 2018).

Importantly, remote warfare is not carried out solely via remote weapons systems, which is sometimes dubbed 'remote control war' (Gusterson 2016). Remote technologies play a role, but remote warfare encompasses a broader set of actions. Ultimately, the activities which make up remote warfare are undertaken to counter an adversary, which often takes the form of non-state armed groups (Knowles and Watson 2018).

Remote warfare normally involves states using and combining the following measures:

- Supporting local security forces, either official state forces, militias or paramilitaries; for example, through the provision of training, equipment or both
- Special operations forces, either training or sometimes even working alongside local and national forces
- Private military and security contractors undertaking a variety of roles (which are discussed in greater detail in the chapter by Christopher Kinsey and Helene Olsen)
- Air strikes and air support, including unmanned aerial vehicles (UAVs) or 'armed drones' and manned aircraft
- Sharing intelligence with state and non-state partners involved in frontline combat (explored by Julian Richards' chapter)

How and where is it being used?

There are several instances where states have shied away from deploying large numbers of 'boots on the ground' and opted for remote approaches. The 2011 NATO-led intervention in Libya is an illustrative case of this. With the desire to avoid the costly consequences of occupation seen in Iraq and Afghanistan, the Obama administration and its international allies supported Libyans to do the bulk of the fighting against Muammar Gaddafi. Faced with what, at the time, seemed to be a looming humanitarian crisis, UN Resolution 1973 was passed and called for the protection of civilians against threats by the Gaddafi regime. Initially, this was confined to several air strikes (see Mueller 2015). But it shifted to small numbers of boots being deployed on the ground (for good overviews see O'Hanlon 2011; Chesterman 2013; Murray 2013; Engelbrekt, Mohlin and Wagnsson 2013). Despite the initial goal of protecting civilians, the intervention became focussed on regime change. French, British and Qatari special forces were sent to assist and train the Libyan rebels and intelligence assets were used to support the rebels as they advanced (Mueller 2015). Overall, the use of remote warfare was crucial in overthrowing Gaddafi. But as explored later, Libya is a compelling example of some of remote warfare's serious problems.

Another salient example of remote warfare in practice is the US-led coalition's support to local forces in Iraq and Syria to counter the Islamic State (commonly known as ISIS) in Iraq and Syria. In Syria, the US trained and equipped units of the Free Syrian Army and Harakat al-Hazm. In northern Iraq, US Special Forces and others trained and supported Peshmerga fighters. Air strikes were conducted heavily throughout these campaigns to support efforts on the ground (Airwars 2016). These actions were undertaken with minimal financial and human costs for the Western militaries involved and, for the most part, successfully pushed back ISIS (Krieg 2016, 109). But the lack of long-term planning for the post-ISIS phase of the conflict has had grave and lasting consequences.

The activities which make up remote warfare are often, but not always, undertaken in secret. Though they can attract media attention, remote warfare's engagements are largely kept out of the public eye. They are often part of 'grey zone conflicts', which describes hostile and aggressive activities that remain 'above and below' the threshold of what is perceived as war (Carment and Belo 2018).

Remote warfare's generally opaque character makes it difficult to gain a complete picture of its use around the globe. But its presence is discernible in many continents. It can be seen in counter-terrorism campaigns in the Middle East, the Horn of Africa, the Sahel and South-East Asia. It is also part of efforts to address near-peer threats, with many states developing a strategy of 'persistent engagement' which sees small numbers of forces around the world working with local partners to build influence and local knowledge to gain an edge over their adversaries (Watson 2020a).

A Western way of war?

Remote warfare has come to define the Western style of military engagement in the first quarter of this century. The US has certainly led the way on this, and many other Western states have followed suit. For example, in the 2015 Strategy and Strategic Defence and Security Review (SDSR), the UK Government pledged to double investment in UK Special Forces and to double the size of the armed drone fleet (HM Government 2015). The UK has also developed a new approach to responding to countries affected by conflict, which includes an increased focus on security sector reform. This activity now makes up a third of the Conflict, Stability and Security Fund's (CSSF) spending (DFID 2019, 28).

Elsewhere in Europe, France, Germany, Italy and even several smaller European states such as the Netherlands, Belgium and Denmark have turned to remote warfare. For instance, several of these states have trained local forces in parts of Africa and the Middle East and conducted air strikes as part of the anti-ISIS Coalition (McInnes 2016). Outside Europe, Australia provided aerial refuelling for the Coalition, shared intelligence and helped train and arm local forces in the fight against ISIS (Airwars 2015, 32).

However, this trend of remote military engagement is not confined to 'the West.' Russia used an assemblage of remote approaches, including special operations forces, military advisers, private militaries, intelligence sharing and local militias to annex Crimea and parts of eastern Ukraine in 2014 (Galeotti 2016). In Syria, the Russians have used a mix of remote methods to stabilise the Assad regime. The Russians also have light footprints in Libya, Venezuela, Mozambique, and the Central African Republic (Kuzio and D'Anieri 2018; Ng and Rumer 2019). Elsewhere, the Iranians have for some time worked with local forces to pursue national objectives across the Middle East through the Islamic Revolutionary Guards Corps' Quds Force (Krieg and Rickli 2019, 164–193). Saudi Arabia, Qatar, Bahrain, Jordan, Turkey and the United Arab Emirates have all used local proxies to counter regional threats (Rondeaux and Sherman 2019). Some African states have a long history of using regional militias to counter non-state armed groups (Craig 2012) and more recently they have employed the services of PMSCs to do this (Varin 2018). Across the globe, then, there is a discernible trend of states engaging militarily from a distance.

A new way of warfare?

Though there are nuances between accounts, several writers have used different terms to describe this type of military engagement. Some of these expressions include 'surrogate war' (Krieg and Rickli 2018), 'risk transfer war' (Shaw 2005, 1), 'vicarious war' (Waldman 2018), 'liquid warfare' (Demmers and Gould 2018), 'network war' (Duffield 2002), 'coalition proxy war' (Mumford 2013), 'postmodern warfare' (Ehrhart 2017) and 'transnational shadow wars' (Niva 2013). Demmers and Gould (2018) have described these terms as attempts to capture the '"new newness" of interventionist warfare'. But there are questions about whether this approach means warfare 'has entered a new era, significantly different from what we have known in the past' (Gat 2011, 28).

Analyses of remote warfare, or other expressions of the phenomenon listed above, are often framed with reference to the 'changing character of warfare' debate. This long-running discourse and associated research enterprise has been trying to 'identify whether war is changing, and – if it is – how those changes affect international relations' (Strachan 2006, 1).

The *character* of war should not be confused with the *nature* of war. The character of warfare is understood, in simple terms, as the ways in which wars are fought. The nature of war, on the other hand, refers to war's enduring essence – or what it is. There is some consensus with conflict researchers and historians that the nature of war has not changed. If we understand warfare as a violent contest of wills between parties (Clausewitz 1832, 1940), then this is seen to still hold true in remote forms of engagement. Nevertheless, the dawn of new technologies, such as artificial intelligence, does represent a significant challenge to the human element of warfare (Johnson 2011; Allen and Chan 2017). Christopher Coker (2002) even speculated that in the future we may witness 'post-human warfare' where machines have replaced humans on the battlefield.

There is a strong case to be made that it is the *character,* rather than the *nature,* of warfare that has changed through the use of remote warfare. From a certain point of view, remote warfare challenges traditional understandings of battlefields and soldiers. 'Intervening' states are now far from the frontlines, providing training in fortified bases or support from the air through technology. Indeed, the technological leaps seen in the most recent revolution in military affairs has provided the means for states to wage warfare from a distance. Arguments have been made that the use of remote warfare has caused a 'temporal and spatial reconfiguration of war' (Demmers and Gould 2018). From this perspective, the lines between war and peace are seen to have become blurred, because there are now often few clear-cut declarations of war, and the geographical borders and legal frameworks that define conflicts have become hard to discern (Gregory 2010; Banasik 2016; Ehrhart 2017). Scholars have noted that we now see environments in 'fragile states' where there are perpetual conditions of conflict, sometimes named 'forever wars' (Filkins 2009), and shifting mosaics of actors involved with conflicting goals (Badescu 2018).

Yet although the environments of conflicts may be shifting and military technologies evolving, it is noticeable that many of the facets of remote warfare are not necessarily new (Moran 2014, 2–4). The training and arming of local forces by external powers, for example, has been used since antiquity (Williams 2012, 61–63; Krieg and Rickli 2019, 16–18). During the Cold War, the superpowers regularly competed with one another by using locally trained and equipped forces (Mumford 2013). This practice continued well after the fall of the Berlin Wall. Research has found that from 1945 to 2011, external actors provided explicit or alleged support to 48 percent of 443 rebel groups engaged in armed conflict (Cunningham, Gleditsch and Salehyan 2013).

Another aspect of remote warfare, the private military and security industry, emerged in the 1980s and began to play a significant role in global security affairs in the 1990s (Krieg 2018, 1). Today, it is a global industry estimated to be worth somewhere between £69 billion and £275 billion a year (Norton-Taylor 2016). Governments are some of the biggest contemporary clients and have found considerable use for the services offered by security contractors (see Kinsey 2006). In 2012, *The Economist* reported that the US Government had 20,000 contractors in Iraq and Afghanistan alone (The Economist 2012). But, as the chapter in this volume by Christopher Kinsey and Helene Olsen shows, this trend of states using 'mercenaries' has been charted back to as early as the sixteenth century and possibly before (see also Parrot 2012).

In many countries, especially the US, special forces 'have grown in every possible way – from their budget to their size, to their pace of operations, to the geographic sweep of their missions' (Turse 2018). Yet despite their recent global proliferation, special forces' origins, at least conceptually, are often seen to lie with the use of the Desert Rats in the First World War (Moreman 2007). They were officially established as part of the British military in the Second World War, with the Special Air Service (SAS) (see Finlan 2009; Karlshøj-Pedersen 2020).

Even the use of UAVs as an instrument of armed conflict is not necessarily as new as some might think. Hugh Gusterson (2016) documents how the first 'armed drone' aircrafts were developed in the First World War as crude radio-controlled biplanes intended to be bombers. It is true that their regular usage has only been in effect for the last decade and a half and this continues to proliferate. For example, they have now become a method employed by non-state actors (Abbot, Clarke and Hathom 2016). Nevertheless, UAVs were used in some form in twentieth-century conflicts, including the Vietnam War (1955–1975), the Yom Kippur War (1973), the Gulf War (1990–1991) and the NATO intervention in Kosovo (1998–1999) (Chamayou 2007, 28).

There are historical examples of states combining some of the methods associated with remote warfare while maintaining a degree of distance from the frontline. The British Empire used local authorities and military auxiliaries, as well as technological tools such as airpower as a form of 'colonial policing' (see Omissi 2017; Marshall 2016). The US employed several approaches associated with remote warfare – such as support for paramilitaries and intelligence sharing – in the Cold War as part of its covert activities in Latin America and elsewhere (Grow 2018; O'Rourke 2018). More recently, in the NATO-led Kosovo campaign at the end of the nineties, Western forces did not deploy large numbers of their own troops and instead used air strikes to support regional troops (see Ignatieff 1998, 169). As Jonathan Gilmore has argued on Kosovo 'there were indicators of a desire amongst Western interveners to have less skin in the game' (Oxford Research Group 2018).

Before 9/11, Donald Rumsfeld believed the US would counter threats in the post-Cold War world with the 'use of airpower, special forces and expeditionary units rather than boots on the ground' (see Rogers 2012).

Elements of this 'Rumsfeld Doctrine' were seen in Afghanistan in late 2001 where a combination of sustained air attacks, deployment of special forces and Central Intelligence Agency (CIA) operatives, and strategic support of the Northern Alliance warlords was used to overthrow the Taliban (Rogers 2016, 24–35). What we are seeing now, though, is an increasing reliance on remote warfare by states, which has arguably not been seen on this global scale before.

How did we get here?

There are several reasons why states have employed this approach. Focusing on Western democracies, the next chapter by Demmers and Gould explores this in greater detail. But it is worth noting that the Iraq and Afghanistan wars have been important drivers.

These conflicts, which began at the start of the century, never truly ended. Nearly two decades on from their inceptions, the costs of these interventions in lives, money and prospects for peace made many legislatures and publics sceptical about the utility of military force abroad (Gribble et al. 2015; Bilmes 2013; Crawford 2018; Holmes 2020).

By the late 2000s, many political leaders promised the end of heavy military interventions and withdrew troops from some theatres (BBC 2011). Yet at the same time, these leaders continued to fear the presence of non-state armed groups. The Arab Spring, which began in late 2010, also caused anxiety. With the instability created by the violent reactions to protest movements, analysts warned that non-state armed groups would thrive (Bokhari 2011).

Faced with the dilemma of wanting to confront perceived threats to national and international security against the backdrop of low popular support for military engagement, the Obama administration sought a different approach to large-scale, 'boots on the ground' interventions. The chosen path was a shift to light-footprint methods (Goldsmith and Waxman 2016, 8-9; Goldberg 2016).

In 2012, following a major strategic review of US security, then-President Obama formally declared 'the end of long-term nation-building with large military footprints' and a move towards 'innovative, low-cost, small-footprint approaches' to achieve America's security objectives (Obama 2012). In light of this, America's general preference in the era of 'Iraq and Afghanistan syndrome' (K.P. Mueller 2005 and 2011) has been to fight its wars by supporting local, national and regional forces and limiting the exposure of its own military to harm.

Concerns about public war-weariness also seem to be an important driver behind the UK's decision to use remote warfare. A leaked Ministry of Defence document from 2013 suggested how to maintain military operations despite a 'risk-averse' public (Quinn 2013).

For risk-averse politicians, then, the use of remote warfare is appealing. It appears to get around military, political and economic restrictions by removing a country's own forces from the frontline. This minimises the scrutiny over military engagements abroad. It allows states to deny responsibility because they are often not directly involved in combat operations or their special forces operations are clouded in secrecy. But there are significant problems with this use of remote warfare. The next section explores this in more depth, focusing mainly on the UK's use of the practice.

The perils of remote warfare: some observations from the British experience

While remote warfare has become increasingly relied upon by the UK, research conducted on its use over the last six years by ORG has shown that it carries significant risks. It often shifts the burden of risk onto civilians; exacerbates the drivers of conflict; and undermines democratic oversight on the use of force abroad. These problems are interconnected.

Protection of civilians

The fact that states like the UK intervene on a light footprint does not mean that the risks of military intervention are removed, or even mitigated against (Knowles and Watson, 2018a). In fact, by shifting the burden of responsibility to partner forces, the UK is increasing the risks to civilian populations because they support partners who may lack the capabilities, willingness or training to sufficiently protect civilians (Walpole and Karlshøj-Pedersen 2019, 2020).

The anti-ISIS coalition's activities in Iraq and Syria highlight this clearly. For example, in both Raqqa and Mosul, where the anti-ISIS coalition was assisting the Iraqi Security Forces and Syrian Democratic Forces (SDF) respectively, 'the coalition largely sat back and provided fire support' in the form of artillery and air strikes to uproot ISIS fighters who had 'years to prepare defensive positions' (Rempfer 2019). This strong reliance on air support for a partner force, which proved unable to implement strong protection of civilian mechanisms, had dramatic consequences for the cities of Mosul and Raqqa.

In western Mosul, for instance, around 15 neighbourhoods were destroyed. These districts previously housed around 230,000 residents, leaving large numbers of internally displaced people who will not be able to return in the short to mid-term (UN News 2017). Three-quarters of Mosul's roads, all of its bridges, and most of the electrical network were also destroyed, and many buildings rigged with explosives and booby traps by retreating ISIS fighters (Kossov 2017). UN estimates suggest that 8 out of 10 buildings damaged in Mosul were residential buildings, with 8,475 houses destroyed – more than 5,500 of which in west Mosul's Old City (Rodgers, Stylianou and Dunford 2017).

British Major General Rupert Jones, who was part of the US-led anti-ISIS coalition, made the following observation when giving oral evidence to the Defence Select Committee in the British parliament:

> I don't think any military in living memory has encountered a battle of this nature. I have said regularly – I stand ready to challenge – that I cannot think of a more significant urban battle since the Second World War (Jones 2018).

In Raqqa, despite being described by US General Stephen Townsend as 'the most precise campaign in the history of warfare' (US Department of Defense 2017), the bombardment left eighty percent of the city destroyed and more than 11,000 buildings uninhabitable (Amnesty 2018).

Ultimately, remote warfare makes the tracking of civilian casualties difficult. Western countries have less capacity to place their troops on the frontlines to carry out the same level of pre- and post-strike assessments that proved to be crucial for reducing civilian casualties in the Afghan theatre (Walpole and Karlshøj-Pedersen 2019). Relying exclusively on ISTAR (intelligence, surveillance, target acquisition and reconnaissance) from UAVs to track civilian harm, as it is so often done now, is ineffective because this approach cannot provide the eyes and ears on the ground needed to conduct thorough investigations (Ibid.). Overall, the UK Government has shown a lack of will to either acknowledge the distinct risks to civilians in these recent military campaigns or to adapt its approach to tracking civilian harm (Walpole and Karlshøj-Pedersen 2020).

Long-term drivers of instability and conflict

Remote warfare also risks exacerbating, rather than resolving, the drivers of conflict. Most of the problems in the places where the UK is engaged are deeply political and require political solutions. Yet remote warfare tends to be short-term and militarily focussed (Knowles and Watson 2018b). So, when the users of remote warfare fail to properly check the background of prospective partners, as they often do, they risk making matters worse by building the capacity of predatory, sectarian or unrepresentative armed groups or national militaries. This can prolong violent conflicts and help create the 'forever wars' that have come to define today's international security environment (Watson and Karlshøj-Pedersen 2019).

Libya is a notable example of this. Following Gaddafi's fall, the country descended into chaos. Within this disorder, ISIS took Sirte and expanded its presence into several surrounding towns and villages. In response, some Western countries, such as France, the US and the UK, engaged in a second, though underreported, round of remote warfare to push ISIS from the country. Part of this process involved empowering non-state groups, including militia from Misrata and the Libyan National Army led by the controversial Khalifa Haftar. At the same time as providing military support to these groups, the West publicly supported the UN-backed Government of National Accord (GNA). While some of these groups were nominally aligned with the GNA, the Government had no meaningful control over them. In fact, Fayez al-Sarraj, the Prime Minister of the GNA, lamented in November 2016: 'They do as they please [...] Whenever they want to go out and fight, they don't ask us and we end up firefighting these battles' (quoted in Zaptia 2016). So, by supporting these groups, the US, France and the UK undermined the GNA's legitimacy and strengthened direct threats to its authority – to the detriment of peace and stability (Watson 2020b forthcoming). Despite pushing ISIS back, Libya remained polarised and fragmented (Wehrey and Lacher 2017).

Field research in post-Gaddafi Libya by Alison Pargeter (2017, 3) noted that the international approach had 'alter[ed] the balance of power on the ground, which has the potential to further undermine the prospects for peace.' This diagnosis was proved correct when, in April 2019, just a few days before the UN was due to hold a conference to establish Libya's 'path to reconciliation and elections', Haftar's forces launched a military campaign, named 'Flood of Dignity' to take Tripoli from the GNA (Trauthig 2019).

The fighting in Libya is still ongoing, but it is a different conflict to one which started nearly a decade ago, with international actors now backing various sides (Allahoum 2020) and various armed factions competing for control over land and resources (Megresi 2019). The conflict between the LNA and the Tripoli-based government is made even more complicated and protracted by the involvement of external actors such as Russia, Egypt, France and the United Arab Emirates. Similarly, third-party security contractors and mercenaries have played an increasingly important role in the conflict (Vest and Clare 2020; Lacher 2019). The UN recently warned that this has contributed to the escalation of the conflict in Libya (OGHCR 2020).

Libya's chronic instability has had huge implications for civilians. A UN Official recently remarked that the impact of Libya's nine-year war on civilians is 'incalculable', with rising casualties and nearly 900,000 people now needing assistance (Lederer 2020). Despite UN mediation efforts, the conflict shows no signs of being resolved soon.

Libya's plight is by no means the only example of how remote warfare contributes to instability and prolongs conflict. The Western footprint during the anti-ISIS coalition in Iraq and Syria was small because states relied on local groups. Some of these have real or perceived ethnic, geographical or community bias, such as the Peshmerga in Iraq and the SDF in Syria. This has undermined the legitimacy of these groups among local and regional actors (Knowles and Watson 2018a). By working with them, international forces exacerbated local, regional, and international tensions and, arguably, created more fragmentation and instability in the future.

In Iraq, empowering the Peshmerga throughout the campaign now threatens to weaken the unity of an already fragmented Iraqi security sector (Knowles 2018a). Now, many Iraqis claim that the Iraqi Army 'is lucky if it can be considered the fourth-strongest army in Iraq – behind, Kurdistan's Peshmerga forces, the Popular Mobilisation Forces and Iraqi tribal fighters' (Mansour and al-Jabbar 2017).

In Syria, working with the SDF pushed back ISIS and established enduring governance structures in Kurdish majority areas, but it was not seen as legitimate by Arab communities (Watson 2018a). Moreover, the perceived links between the SDF and the Kurdish Workers Party (PKK) – a group leading an armed insurgency against the Turkish state – has meant that support to the group remains unacceptable to the Turkish government (Watson 2020). This has worsened relations between the West and NATO ally Turkey.

Serious problems are also evident when states provide support to other state forces (Watson 2020b). The Armed Conflict Location and Event Data Project noted:

> Governments continue to pose the greatest threat to civilians
> around the world, with state forces responsible for more than a
> quarter of all violence targeting civilians in 2019 – the largest
> proportion of any actor type. Of the top five actors responsible

for the largest share of civilian targeting in 2019, four of them are state forces, and the fifth is a progovernment militia. (Kishi, Pavlik and Jones 2020)

Given this trend, the international community should not respond to instability by providing light-footprint training (which is militarily and technically focussed) for national armies (Kleinfeld 2019). Yet, they often do; the US Stabilization Assistance Review noted:

In support of counterterrorism objectives, the international community is providing high volumes of security sector training and assistance to many conflict affected countries, but our programs are largely disconnected from a political strategy writ large, and do not address the civilian military aspects required for transitional public and citizen security. (Bureau of Conflict and Stabilization Operations 2018)

Activities that focus on 'defence and security institutions' but allow oversight to remain 'weak and ineffective [...] can lead to a situation where rights-violating security forces become better equipped to do what they have always done' (Caparini and Cole 2008). Many governments in Africa, the Middle East and elsewhere have used international support to increase the capacity of their security sectors but have failed to address corruption and abuses by predatory state forces (Transparency International 2019). This 'risk[s] further undermining human security' when populations are trapped 'between increased violence of abusive security forces and the terror of non-state armed groups' (Knowles and Matisek 2019). This, in turn, risks further alienating the civilian population and pushing them towards extremist groups (Watson and Karlshøj-Pedersen 2019). In Somalia, field research found that that the abuses of the Somali National Army are 'a big recruitment tool for Al Shabab' (Knowles 2018b). Similarly, an International Alert study on young Fulani people in the regions of Mopti (Mali), Sahel (Burkina Faso) and Tillabéri (Niger) found 'real or perceived state abuse is the number one factor behind young people's decision to join violent extremist groups' (Raineri 2018, 7).

Transparency and accountability

Efforts to address these risks are undermined by the poor transparency and accountability of remote warfare. Over the last two decades in the UK, there has been an increased recognition that the decision to use force abroad should not sit solely with the Prime Minister (Knowles and Watson 2017). This recognition drove the development of the War Powers Convention, a constitutional convention mandating the House of Commons to accept or reject proposed deployments of the British Armed Forces on combat operations abroad (Walpole 2017).

However, in many ways, remote warfare falls through the gaps in mechanisms designed to oversee the use of force abroad. Train and assist operations are often not designated as 'combat missions' (even when they are in contested areas or close to the frontline) and so do not necessarily fall under the War Powers Convention (Karlshøj-Pedersen 2018b). This is despite the fact there is no official definition of combat and non-combat operations or a set list of criteria (Blunt 2018). Further, Ministry of Defence and CSSF annual reports discuss these activities but usually only release headlines for some programmes and are inconsistent year to year (Karlshøj-Pedersen 2018a).

Like many states, the UK has seen the number and remit of its special forces increase since 9/11 (Moran 2016, 3–5). The ease with which prime ministers can deploy special forces, without recourse to Parliament, has increased the appeal of their use. This sees them increasingly deployed not just in support of conventional forces, but also as 'instruments of national power' in many parts of the world today. Despite these developments, UK Special Forces have continued to lack sufficient scrutiny because of the government's long-held blanket opacity policy that precludes any form of external oversight (Walpole and Karlshøj-Pedersen 2018). While committees have a long history of overseeing British action abroad, including the actions of the secretive intelligence agencies, they are unable to scrutinise the actions of Special Forces and information about their use is specifically exempt from the Freedom of Information Act (Ibid.). Special Forces are the only piece of the UK's defence, security, and intelligence apparatus to continue to fall outside of any parliamentary oversight. It has long been accepted that 'the MoD's long-held policy […] is not to comment on Special Forces' (Knowles 2016). As Earl Howe, a Conservative House of Lords front-bencher, remarked in 2018, 'It is this Government's, and previous Governments', policy not to comment, and to dissuade others from commenting or speculating, about the operational activities of Special Forces' (UK Parliament 2018).

This deniability around the use of UK Special Forces may bring flexibility, which creates opportunities when it comes to dealing with the fluid and complex security threats animating today's global security landscape. But this is not a simple relationship whereby more secrecy automatically brings greater strategic advantages. As noted, the prevailing tendency towards secrecy is creating an accountability gap that challenges the UK's democratic controls over the use of force. In addition to being democratically precarious, it restricts the government's ability to set its own narrative for British military action overseas. Shaping the narrative around conflicts has always been important for parties, but the growing interconnectedness that the information age brings has elevated the significance of this in military and political debates (Knowles and Watson 2017, 5). The 2010 SDSR made this point very clearly when it said 'the growth of communications technology will increase our enemies' ability to influence, not only all those on the battlefield, but also our own society directly. We must therefore win the battle for information, as well as the battle on the ground' (HM Government 2010, 16). However, secretive policies risk 'exacerbating the low levels of public trust in government' and preventing the UK from effectively shaping public narratives (Knowles and Watson 2018, 28).

The frequent reports of UK Special Forces in the media have created an uneasy coexistence of official opacity and sporadic leaks of information (Knowles and Watson 2017). This has led to discrepancies between official statements and media revelations. Such media reports include the 2011 incident in which an SAS team was arrested by Libyan rebels (Jabar 2011), the BBC's 2016 publication of images showing SAS forces fighting in Syria (Sommerville 2016), the reports in 2019 that British troops had been fighting alongside a Saudi-funded militia in Yemen who allegedly recruited child soldiers (Wintour 2019), and the recent allegations of UK Special Forces executing unarmed civilians in Afghanistan (Arbuthnott, Calvet and Collins 2020).

Furthermore, the shroud of secrecy that covers UK Special Forces operations means it is unclear how consistently strategic concerns about their impact on long-term stability are factored into decision-making around their use. UK Special Forces are not immune to such dangers, especially if they are often engaged in more kinetic activities than regular soldiers. The Foreign Affairs Committee made the following comments in 2016 when it emerged that UKSF had been on the ground in Libya:

> Special Forces operations in Libya are problematic because they necessarily involve supporting individual militias associated with the [UN-backed Government of National Accord] rather than the GNA itself, which does not directly command units on the ground […] Special Forces missions are not currently subject to parliamentary or public scrutiny, which increases the danger that such operations can become detached from political objectives. (House of Commons Foreign Affairs Committee 2016)

A lack of oversight, then, does not necessarily make UK Special Forces more effective. Instead, the fact that none of these concerns could be alleviated may mean that fatal assumptions and bad strategy are not properly checked. The blanket opacity also makes it impossible to assess the effectiveness of their approach to civilian harm mitigation. When operations go wrong and civilians are harmed it is unclear whether lessons are being learned and steps being taken to avoid the same mistakes from recurring. There is further uncertainty over whether there are adequate processes in place to ensure allegations of wrongdoing are met with the same due process which applies to the rest of Britain's Armed Forces (Walpole and Karlshøj-Pedersen 2020). A failure to promptly and adequately hold UK forces to account for transgressions is likely to have serious reputational consequences with both its international allies and local populations in the theatres where the UK is engaged (Ibid.).

In a world of smartphones, social media and burgeoning access to the internet, controlling the flow of information on UK military action abroad and keeping special operations secret – including scandals around their involvement in civilian harm – has become even harder (Knowles and Watson 2017). These realities make the culture of no comment remarkably outdated.

The UK's approach to special forces oversight contrasts heavily to many of its allies. Some countries – the US, France, Denmark and Norway – have adopted some form of legislative scrutiny, with Denmark's system being the most expansive and France's the most limited (Walpole and Karlshøj-Pedersen 2018, 18). Others, Australia and Canada, have adopted a policy of releasing unclassified briefings on the activities of their special forces, which can then be used by the media, the public, and their legislatures as a basis for debate (Ibid.).

Even when it comes to British involvement in the US-led air campaign against ISIS, which was approved by a parliamentary vote, discussion of the UK's impact has remained poor (Watson 2018b). For instance, while the Ministry of Defence claims to have killed or injured 4000 ISIS fighters, they have only admitted to killing one civilian (Knowles and Watson 2018b). This account has been proven to be implausible by several studies (Amnesty International 2019; Walpole and Karlshøj-Pedersen 2020). In Mosul, for instance, of the 6,000 to 9,000 alleged civilian deaths estimates suggest that between 1,066 and 1,579 of those deaths were caused by Coalition actions (Airwars 2018, 7). In Raqqa, investigations suggested at least 1,400 civilian fatalities could be tied to Coalition activities (Ibid., 8).

The lack of transparency around the UK's remote warfare leads to ineffective accountability, with adverse consequences for the protection of civilians. A significant scrutiny gap means that the government does not understand the short- and long-term impact of its operations. These interlinked problems can help perpetuate the cycles of violence seen in the many theatres of remote warfare.

Conclusion

In an era where there is a greater emphasis placed on state-on-state competition, remote warfare seems to be here to stay. Yet, many states – evidenced in doctrines, budgets and practical deployments – show a future commitment to light-footprint interventions even with this rise of great power competition. It is troubling that such developments look likely to continue given that there is little appreciation of the political, ethical and legal implications. This makes a broader debate about the risks of this type of intervention essential going forwards. The remaining chapters in this book bring together a range of experts from various backgrounds who provide a deeper dive into the pitfalls of remote warfare.

References

Abbott, Chris, Matthew Clarke, Steve Hathorn and Scott Hickie. 2016. *Hostile drones: the hostile use of drones by non-state actors against British targets*. Open Briefing Report.

Airwars. 2015. *Cause for Concern – Hundreds of civilian non-combatants credibly reported killed in first year of Coalition air strikes against Islamic State*. Airwars Report. https://airwars.org/report/cause-for-concern-hundreds-of-civilian-non-combatants-credibly-reported-killed-in-first-year-of-coalition- air strikes-against-islamic-state/

———. 2016. *Limited Accountability: A transparency audit of the anti-ISIL Coalition*. Oxford Research Group.

———. 2018. *Death in the City – High levels of civilian harm in modern urban warfare resulting from significant explosive weapons use*. Airwars Report.

Allahoum, Ramy. 2020. 'Libya's war: Who is supporting whom.' *Al Jazeera*. 9 July. https://www.aljazeera.com/news/2020/01/libya-war-supporting-200104110325735.html

Allen, Greg, and Taniel Chan. 2017. *Artificial intelligence and national security*. Cambridge, MA: Belfer Center for Science and International Affairs.

Amnesty International. 2018. '*War of annihilation': Devastating toll on civilians, Raqqa – Syria*. https://www.amnesty.org/en/documents/mde24/8367/2018/en/

————. 2019 'Our Investigation Reveals US-Led Coalition Killed Thousands of Civilians in Raqqa.' Amnesty International News. April. https://www.amnesty.org/en/latest/news/2019/04/syria-unprecedented-investigation-reveals-us-ledcoalition-killed-more-than-1600-civiliansin-raqqa-death-trap/

Arbuthnott, George, Jonathan Calvert and David Collins. 2020. 'Rogue SAS Afghanistan execution squad exposed by email trail.' *The Times*. 1 August.

Banasik, Miroslaw. 2016. 'Unconventional war and warfare in the gray zone. The new spectrum of modern conflicts.' *Journal of Defense Resources Management (JoDRM)*, 7(1): 37–46.

BBC News. 2011. 'US President Barack Obama announces Iraq withdrawal.' 21 October. https://www.bbc.co.uk/news/av/world-us-canada-15410534

Bilmes, Linda. 2013. 'The Financial Legacy of Iraq and Afghanistan: How Wartime Spending Decisions Will Constrain Future National Security Budgets.' HKS Faculty Research Working Paper Series RWP13-006.

Blunt, Crispin. 2018. 'In search of clarity: defining British combat and non-combat operations.'1828. https://www.1828.org.uk/2018/09/28/in-search-of-clarity-defining-british-combat-and-non-combat-operations-2/

Bokhari, Kamran. 2012. 'Jihadist Opportunities in Syria.' *Stratfor*. 14 February.

Bureau of Conflict and Stabilization Operations. 2018. *Stabilization Assistance Review: A Framework for Maximizing the Effectiveness of US Government Efforts To Stabilize Conflict-Affected Areas*. US Department of State.

Caparini, Marina, and Eden Cole. 2008. 'The case for public oversight of the security sector: Concepts and strategies.' *Public Oversight of the Security Sector: A Handbook for Civil Society Organizations*. UNDP.

Carment, David, and Dani Belo. 2018. *War's Future: The Risks and Rewards of Grey Zone Conflict and Hybrid Warfare*. Canadian Global Affairs Institute.

Chalmers, Malcolm, and Will Jessett. 2020. 'Will Defence and the Integrated Review: A Testing Time' *Whitehall Reports*. RUSI. 26 March.

Chamayou, Grégoire. 2015. *Drone Theory*. London: Penguin UK.

Chesterman, Simon. 2011. '"Leading from Behind": The Responsibility to Protect, the Obama Doctrine, and Humanitarian Intervention after Libya.' *Ethics and international affairs*, 25: 279-285.

Clausewitz, von. Carl 1940. [1832]. *On War*. Translated by J.J. Graham. London: Keagan Paul, French, Truebner.

Coker, Christopher. 2013. *Warrior geeks: how 21st century technology is changing the way we fight and think about war*. Hurst Publishing.

Craig, Dylan. 2012. *Proxy war by African states, 1950–2010*. American University.

Crawford, Neta. 2018. 'Human Cost of the Post-9/11 Wars: Lethality and the Need for Transparency.' Brown University, Watson Institute for International and Public Policy.

Cunningham, David, Kristian Skrede Gleditsch and Idean Salehyan. 2013. 'Non-state actors in civil wars: a new dataset.' *Conflict Management and Peace Science,* 30(5): 516–531.

Davidson, Jason. 2013. 'France, Britain and the intervention in Libya: an integrated analysis.' *Cambridge Review of International Affairs*, 26(2): 310–329.

Demmers, Jolle, and Lauren Gould. 2018. 'An assemblage approach to liquid warfare: AFRICOM and the "hunt" for Joseph Kony.' *Security Dialogue*, 49(5): 364–381.

Department for International Development (DFID). 2019. *Conflict, Stability and Security Fund: Annual Report 2018/19*. Her Majesty's Government.

Department of Defense. 2018. Summary of the 2018 National Defence Strategy of the United States. https://dod.defense.gov/Portals/1/Documents/pubs/2018-National-Defense-Strategy-Summary.pdf

Department for International Development (DFID). 2019. *Conflict, Stability and Security Fund: Annual Report 2018/19*. Her Majesty's Government.

Dimitriu, Gabriel. 2020. 'Clausewitz and the politics of war: A contemporary theory.' *Journal of Strategic Studies*, 43(5): 645–685.

Duffield, Mark. 2002. 'Social reconstruction and the radicalization of development: Aid as a relation of global liberal governance.' *Development and Change,* 33(5): 1049–1071.

Ehrhart, Hans-Georg. 2017. 'Postmodern warfare and the blurred boundaries between war and peace.' *Defense and Security Analysis*, 33(3): 263–275.

Engelbrekt, Kjell, Marcus Mohlin, and Charlotte Wagnsson, eds. 2013. *The NATO intervention in Libya: lessons learned from the campaign*. Routledge.

Filkins, Dexter. 2009. *The Forever War*. Vintage.

Finlan, Alistair. 2009. *Special Forces, Strategy and the War on Terror: Warfare by Other Means*. Routledge.

Galeotti, Mark., 2016. 'Hybrid, ambiguous, and non-linear? How new is Russia's "new way of war"'? *Small Wars and Insurgencies*, 27(2): 282–301.

Gat, Azar. 2011. 'The Changing Character of War.' In *The Changing Character of War*, edited by Hew Strachan and Sibylle Scheipers. Oxford: Oxford University Press: 27–47.

Goldberg, Jonah. 2016. 'The Obama Doctrine.' *The Atlantic*, 317(3): 70–90.

Goldsmith, Jack and Matthew Waxman. 2016. 'The Legal Legacy of Light-Footprint Warfare.' *The Washington Quarterly*, 39(2): 7-21.

Gribble, Rachael, Simon Wessley, Susan Klein, David A. Alexander, Christopher Dandeker, and Nicola T. Fear. 2015. 'British public opinion after a decade of war: attitudes to Iraq and Afghanistan.' *Politics,* 35(2): 128–150.

Grow, Michael. 2008. *US Presidents and Latin American Interventions: Pursuing Regime Change in the Cold War*. University Press of Kansas.

Gusterson, Hugh. 2016. *Drone: Remote Control Warfare*. Cambridge, Massachusetts; London, England: The MIT Press.

Holmes, Flora. 2020. *Public Attitudes to Military Intervention*. British Foreign Policy Group.

House of Commons Foreign Affairs Committee. 2016. 'Libya: Examination of intervention and collapse and the UK's future policy options.' *Third report of session* 17.

HM Government. 2015. *National Security Strategy and Strategic Defence and Security Review 2015*. November.

Humud, Carla, Christopher M. Blanchard and Mary Beth Nikitin. 2017. 'Armed Conflict in Syria: Overview and US Response.' *Congressional Research Service Report*. 6 January.

Ignatieff, Michael. 2000. *Virtual War: Kosovo and Beyond*. London: Chatto and Windus: UK.

Jaber, Hala. 2011. 'SAS seized by rebel Libyans.' *The Times*. 6 March.

Johnson, James. 2019. 'Artificial intelligence and future warfare: implications for international security.' *Defense and Security Analysis*, 35(2): 147–169.

Jones, Rupert. 2018. 'Oral Evidence - UK Military Operations in Mosul and Raqqa - 15 May 2018.' UK Parliament.

Karlshøj-Pedersen, Megan. 2018a. 'The True Cost of Defence Engagement?' Oxford Research Group Blog, 14 June. https://www.oxfordresearchgroup.org.uk/blog/the-true-cost-of-defence-engagement

———. 2018b. 'The UK Should Learn from the Transgressions of Australian Special Forces.' *Small Wars Journal*. 14 September.

———. 2018c. 'When is the UK at War?' OpenDemocracy. 20 September.

————. 2020. 'ORG Explains #14: The UK's Special Forces.' Oxford Research Group. 11 May. https://www.oxfordresearchgroup.org.uk/org-explains-14-the-uks-special-forces

Kinsey, Christopher. 2006. *Corporate soldiers and international security: The rise of private military companies*. Routledge.

Kishi, Roudabeh, Melissa Pavlik and Sam Jones. 2020. 'ACLED 2019: The year in review.' Armed Conflict Location and Event Data Project (ACLED).

Kleinfeld, Rachel. 2019. *A Savage Order: How the World's Deadliest Countries Can Forge a Path to Security*. Vintage.

Knowles, Emily. 2016. 'Britain's Culture of No Comment.' Remote Control Project, Oxford Research Group.

————. 2018. 'Falling Short of Security in Somalia.' Oxford Research Group.

Knowles, Emily, and Abigail Watson. 2017. *All Quiet on the ISIS Front: British Secret Warfare in an Information Age*. Remote Control Project, Oxford Research Group.

————. 2018a. *Remote Warfare: Lessons Learned from Contemporary Theatres*. Remote Warfare Programme, Oxford Research Group.

————. 2018b. *No Such Thing as a Quick Fix: The Aspiration–Capabilities Gap in British Remote Warfare*. Remote Warfare Programme, Oxford Research Group.

Knowles, Emily, and Jahara Matisek. 2019. 'Western Security Force Assistance in Weak States: Time for a Peacebuilding Approach.' *The RUSI Journal*, 164(3): 10–21.

Kossov, Igor. 2017. 'Mosul is Completely Destroyed.' *The Atlantic*. 10 July.

Krieg, Andreas. 2016. 'Externalizing the burden of war: the Obama Doctrine and US foreign policy in the Middle East.' *International Affairs*, 92(1): 97–113.

————. 2018. *Defining Remote Warfare: The Rise of the Private Military and Security Industry*. Remote Warfare Programme, Oxford Research Group.

Krieg, Andreas and Jean Marc Rickli. 2018. 'Surrogate warfare: the art of war in the 21st century?' *Defence Studies*, 18(2): 113–130.

———. 2019. *Surrogate warfare: the transformation of war in the twenty-first century*. Georgetown University Press.

Kuzio, Taras, and D'Anieri, Paul. 2018. *Sources of Russia's great power politics: Ukraine and the challenge to the European order*. Bristol: E-International Relations.

Lacher, Wolfgang. 2019. 'Who is Fighting Whom in Tripoli? How the 2019 Civil War is Transforming Libya's Military Landscape.' Briefing Paper. Geneva: Small Arms Survey.

Lamothe, Dan. 2014. 'Retiring top Navy SEAL: "We are in the golden age of Special Operations." *Washington Post*. 29 August. https://www.washingtonpost.com/news/checkpoint/wp/2014/08/29/retiring-topnavy-seal-we-are-in-the-golden-age-of-specialoperations-2/

Lederer, Edith. 2020. 'UN: Impact of long Libya war on civilians is "incalculable"'. *Associated Press*. 18 February. https://apnews.com/333fe41a6bab5d9223aa563507733a03

Lewis, Michael W., 2013. 'Drones: Actually the Most Humane Form of Warfare Ever.' *The Atlantic*. 2 August. http://www.theatlantic.com/international/archive/2013/2008/drones-actuallythe-most-humane-form-of-warfare-ever/278746/

Mansour, Renad, and Fālih 'Abd al-Jabbār. 2017. *The Popular Mobilization Forces and Iraq's Future*. Vol. 28. Washington, DC: Carnegie Endowment for International Peace.

Marshall, Alex. 2016. 'From civil war to proxy war: past history and current dilemmas.' *Small Wars and Insurgencies*, 27(2): 183–195.

McInnis, Kathleen J., 2016. 'Coalition contributions to countering the Islamic State.' *Congressional Research Service, the Library of Congress*. August.

Moran, Jon. 2014. *Remote Warfare (RW): developing a framework for evaluating its use*. Remote Control Project.

————. 2016. *Assessing SOF Transparency and Accountability.* Remote Control Project.

Moreman, Tim, and Duncan Anderson. 2007. *Desert Rats: British 8th Army in North Africa 1941–43.* Osprey Publishing.

Mueller, John. 2005. 'The Iraq Syndrome.' *Foreign Affairs.* November/December.

————. 2011. 'The Iraq Syndrome Revisited.' *Foreign Affairs.* 28 March.

Mueller, Karl. P., ed. 2015. *Precision and Purpose: Airpower in the Libyan Civil War.* RAND Corporation.

Mumford, Andrew. 2013. 'Proxy warfare and the future of conflict.' *RUSI Journal,* 158(2): 40–46.

Murray, Donette. 2013. 'Military action but not as we know it: Libya, Syria and the making of an Obama Doctrine.' *Contemporary Politics*, 19(2): 146–166.

Mutschler, Max M., 2016. 'On the road to liquid warfare? Revisiting Zygmunt Bauman's thoughts on liquid modernity in the context of the "new Western way of war".' BICC Working Paper, Bonn International Center for Conversion.

Ng, Nicole, and Eugene Rumer. 2019. 'The West Fears Russia's Hybrid Warfare. They're Missing the Bigger Picture.' Carnegie Endowment for Democracy.

Niva, Steve. 2013. 'Disappearing violence: JSOC and the Pentagon's new cartography of networked warfare.' *Security Dialogue,* 44(3): 185–202.

Norton-Taylor, Richard. 2016. 'Britain is at centre of global mercenary industry, says charity.' *The Guardian.* 3 February. https://www.theguardian.com/business/2016/feb/03/britain-g4s-at-centre-of-global-mercenary-industry-says-charity

Obama, Barack. 2012. 'Remarks by the President on the Defense Strategic Review.' 5 January. http://www.whitehouse.gov/the-pressoffice/2012/01/05/remarks-president-defense-strategic-review

Office of High Commissioner on United Nations Human Rights. 2020. 'Libya: Violations related to mercenary activities must be investigated – UN experts.' 17 June.

Omissi, David E., 2017. *Air power and colonial control: The royal air force 1919–1939*. Manchester University Press.

O'Rourke, Lindsey A., 2018. *Covert Regime Change: America's Secret Cold War*. Cornell University Press.

Oxford Research Group. 2017. 'The Cosmopolitan Military: An Interview with Jonathan Gilmore.' Oxford Research Group Blog. 20 November. https://www.oxfordresearchgroup.org.uk/blog/the-cosmopolitan-military-an-interview-with-jonathan-gilmore

Pargeter, Alison. 2017. *After the fall: Views from the ground of international military intervention in post-Gadhafi Libya*. Remote Control Project, Oxford Research Group. https://www.oxfordresearchgroup.org.uk/after-the-fall-views-from-the-ground-of-international-military-intervention-in-post-gadhafi-libya

Parrott, David. 2012. *The Business of War: Military Enterprise and Military Revolution in Early Modern Europe*. Cambridge: Cambridge University Press.

Quinn, Ben. 2013. 'MoD Study Sets out How to Sell Wars to the Public.' *The Guardian*. 26 September. https://www.theguardian.com/uk-news/2013/sep/26/modstudy-sell-wars-public

Raineri, Luca. 2018. *If Victims Become Perpetrators: Factors Contributing to Vulnerability and Resilience to Violent Extremism in the Central Sahel*. International Alert.

Rempfer, Kyle. 2019. 'Why US Troops "Flattened" Raqqa and Mosul, and Why It May Herald an Era of "Feral City" Warfare.' *Military Times*. 30 April. https://www.militarytimes.com/news/your-military/2019/04/29/why-us-troops-flattened-raqqa-andmosul-and-what-it-means-for-future-fights/

Rodgers, Lucy, Nassos Stylianou and Daniel Dunford. 2017. 'What's Left of Mosul?' *BBC News*. 9 August.

Rogers, Paul. 2012. 'Remote control, a new way of war.' *openDemocracy*. 18 October.

———. 2016. *Irregular War: The New Threat from the Margins*. Bloomsbury Publishing.

Rondeaux, Candice, and David Sterman. 2019. *Twenty-First Century Proxy Warfare: Confronting Strategic Innovation in a Multipolar World*. New America Foundation.

Shaw, Martin. 2005. *The New Western Way of War: Risk-Transfer War and its Crisis in Iraq*. Cambridge: Polity.

Sommerville, Quintin. 2016. 'UK special forces pictured on the ground in Syria.' BBC News. 8 August.

Strachan, Hew. 2011. 'Introduction.' In *The Changing Character of War*, edited by Hew Strachan and Sibylle Scheipers. Oxford: Oxford University Press: 27–47.

The Economist. 2012. 'Private soldiers Bullets for hire.' 17 November.

The White House. 2017. National Security Strategy of the United States of America. https://www.whitehouse.gov/wp-content/uploads/2017/12/NSS-Final-12-18-2017-0905.pdf

Towle, Philip. 1981. 'The strategy of war by proxy.' *The RUSI Journal*, 126(1): 21–26.

Transparency International. 2019. 'Critical Defence Corruption Threatens Security and Stability in Middle East and North Africa.' 25 November.

Trauthig, Inga K. 2019. 'Haftar's Calculus for Libya: What Happened, and What is Next?' ICSR Insight. The International Centre for the Study of Radicalisation.

Turse, Nick. 2018. 'Commandos Sans Frontières The Global Growth of US Special Operations Forces.' Tom Dispatch. 17 July. https://test.portside.org/node/17823/printable/print

UN News. 2017. 'Recovery in Iraq's War-Battered Mosul Is a "tale of Two Cities", UN Country Coordinator Says.' 8 August. https://news.un.org/en/story/2017/08/563022-recovery-iraqs-war-battered-mosul-tale-two-cities-un-country-coordinator-says

US Department of Defense. 2017. 'Remarks by General Townsend in a media availability in Baghdad, Iraq.' 11 July.

Varin, Caroline. 2018. 'Turning the tides of war: The impact of private military and security companies on Nigeria's counterinsurgency against Boko Haram.' *African Security Review*, 27(2): 144–157.

Vest, Nathan and Colin Clarke. 2020. 'Is the Conflict in Libya a Preview of the Future of Warfare?' *Defense One*. 2 June. https://www.defenseone.com/ideas/2020/06/conflict-libya-preview-future-warfare/165807/

Waldman, Thomas. 2018. 'Vicarious warfare: The counterproductive consequences of modern American military practice.' *Contemporary Security Policy*, 39(2): 181–205.

Walpole, Liam. 2017. 'Mind the Gap: Parliament in the Age of Remote Warfare.' Oxford Research Group. 31 October.

Walpole, Liam and Megan Karlshøj-Pedersen. 2018. *Britain's Shadow Army: Policy Options for External Oversight of UK Special Forces*. Oxford Research Group.

———. 2019. 'Remote Warfare and the Practical Challenges for the Protection of Civilians Strategy.' Oxford Research Group.

———. 2020. *Forging a New Path: Prioritising the Protection of Civilians in the UK's Response to Conflict*. Oxford Research Group. https://www.oxfordresearchgroup.org.uk/forging-a-new-path-prioritising-the-protection-of-civilians-in-the-uks-response-to-conflict

Watson, Abigail, and Megan Karlshøj-Pedersen. 2019. *Fusion Doctrine in Five steps: Lessons Learned from Remote Warfare in Africa*. Oxford Research Group.

Watson, Abigail. 2018. 'Pacifism or Pragmatism? The 2013 Parliamentary Vote on Military Action in Syria.' Oxford Research Group.

———.2020a. 'Questions for the Integrated Review #2: how to engage: deep and narrow or wide and shallow?' Oxford Research Group. 14 July.

————. 2020b. 'Uneasy Alliances: The political consequences of arming nonstate groups in counter-terrorism campaigns.' In *Violent Nonstate Actors in Conflict,* edited by David Brown, Murray Donette, Malte Riemann and Norma Rossi. Howgate Publishing.

Watts, Tom, and Rubrick Biegon. 2018. 'Conceptualising Remote Warfare: Past, Present and Future.' Oxford Research Group.

Wehrey, Frederic, and Wolfram Lacher. 2017. 'Libya After ISIS'. *Foreign Affairs*. 22 February.

Williams, Brian Glyn. 2012. 'Fighting with a Double-Edged Sword?: Proxy Militias in Iraq, Afghanistan, Bosnia, and Chechnya.' In *Making Sense of Proxy Wars: States, Surrogates & the Use of Force,* edited by Michael Innes.

Wintour, Patrick. 2019. '"Serious" questions over SAS involvement in Yemen war.' *The Guardian*. 27 March.

Zaptia, Sami. 2016. 'Serraj blames Hafter, Saleh, Elkaber and Ghariani for Libya's problems.' *Libya Herald*. 2 November.

2

The Remote Warfare Paradox: Democracies, Risk Aversion and Military Engagement

JOLLE DEMMERS AND LAUREN GOULD

Liberal Western democracies are increasingly resorting to remote warfare to govern security threats from a distance. From the 2011 NATO-led bombings in Libya, the US Africa Command training Ugandan soldiers to fight al-Shabaab, or the US-led Coalition against Islamic State (ISIS) in Syria and Iraq, violence is exercised from afar. Remote warfare is characterised by a shift away from 'boots on the ground' deployments towards light-footprint military interventions. This may involve using drone and air strikes, special forces, intelligence operatives, private contractors and training teams assisting local forces to do the fighting, killing, and dying on the ground (Demmers and Gould 2018, 365; Watts and Biegon 2017, 1). Violence is thus exercised without exposing Western military personnel to opponents in a declared warzone under the condition of mutual risk. This chapter aims to understand why we see this shift to remote warfare and reviews the moral and political challenges that this new way of war has given rise to. Our key argument is that the secrecy around remote warfare operations, their portrayal as 'precise' and 'surgical', as well the asymmetrical distribution of death and suffering they entail, thwarts democratic political deliberation on contemporary warfare. We foresee that it is these qualities of remote warfare that will make Western liberal democracies more war prone, not less. This is the remote warfare paradox: the military violence executed is rendered so remote and sanitised, that it becomes uncared for, and even ceases to be defined as war.

To empirically illustrate these points, we start with two vignettes. The first highlights how 'war' was taken out of the air strike campaign in Libya in 2011.

Libya and US air strikes in 2011

In 2011, when NATO bombs were dropped on Libya, US President Obama faced a dilemma: would he have to end US air strikes after the sixtieth day, as required by the War Powers Resolution? The War Powers Resolution (WPR) is a 1973 law that orders the president to withdraw US forces from 'hostilities' within 60 days in the absence of congressional authorisation. It was clear at the time that Congress had little interest in supporting the intervention in Libya with a war declaration or statutory authorisation.

In response, White House lawyers crafted a legal rationale that allowed Obama to bypass the WPR predicament. The Libya war did not 'rise to the level of hostilities', they concluded, because military engagement was limited by design, conducted without the involvement of US ground forces, and therefore free of any risk of friendly casualties. The lawyers' report asserted that 'US operations do not involve sustained fighting or active exchanges of fire with hostile forces, nor do they involve the presence of US ground troops, US casualties or a serious threat thereof […].' (US Department of State 2011)

By placing US air strikes outside the realm of hostilities as envisioned by the WPR, it was argued that Obama did not need Congressional authorisation to engage in an offensive mission involving sustained bombardments of a foreign government's forces. With this argument, the Obama administration set a precedent, drawn upon by his successor, that consistent bombing does not rise to the level of 'hostilities' (Olsen 2019). In 2019, the Trump administration was easily able to claim that US support for the Saudi-led coalition in Yemen – merely refuelling jets and intelligence sharing, well short of direct air strikes – does not constitute hostilities.

Syria and the Dutch support for militias in 2015

Our second vignette shows another dimension of remote warfare: here, too, a Western democracy is waging remote war, but here the remoteness entails an 'outsourcing' that is shrouded in secrecy.

In 2015, the Dutch parliament authorised the Ministry of Foreign Affairs to provide non-lethal support to militias fighting ISIS and Assad in Syria. This permission was granted under the conditions that only 'moderate' groups with respect for international humanitarian law would be assisted; support would be terminated immediately if this proved not to be the case, and the parliament would be given frequent updates about the programme.

In 2018, Dutch newspaper *Trouw* revealed that the Netherlands had supported over 22 militias by providing them with 25 million euros worth of pick-up trucks, laptops and uniforms. *Trouw* also exposed that this support had been used for 'lethal' ends and the Ministry of Foreign Affairs was well aware that a number of these militias were human rights abusers (Dahhan and Holdert 2018).

The Dutch Public Prosecutor had even labelled one, Jabhat al-Shamiya, a terrorist organisation (Ibid.). The parliament was shocked by these revelations; however, in response the Ministry of Foreign Affairs labelled the programme as 'classified,' thereby ruling out public, political and legal enquiry.

These vignettes illustrate how removing Western military personnel from the theatres of war through air strikes or the financing of local militias and framing these distant engagements as 'non-hostile', 'non-lethal' or 'classified' offers room for governments to bypass domestic public scrutiny and political debate. Aiming to explore this 'depoliticisation move' in more detail, we do three things in this chapter.

First, in order to set the larger picture, we explore three reasons to explain the turn to remote warfare: the desire for leaders to avoid the risks associated with warfare, the rise of technological developments, and the networked character of modern warfare. Second, we highlight how these debates fail to address the transformative nature of remote warfare, namely that it allows Western states to wage bloodless wars while transferring the risk of death and suffering to local populations. This raises the question of whether Western democracies will *'care'* enough to restrain the wars waged in their name. In the third section, we investigate the answer to this question by highlighting how watchdog and human rights organisations such as Airwars and Amnesty International try to make Western publics and parliaments *'care'* by publicising the local human suffering caused by Western air strikes. In the conclusion, we come back to the 'paradox of remote warfare' and the challenges that remote killings present to the politics of war: both to public scrutiny and political decision-making.

Explaining the turn to remote warfare

Why do we see this turn to 'remote' interventionism? And why now? Aiming to capture the 'new newness' of interventionism, scholars and specialists have coined a range of terms. We see labels popping up such as 'globalized war' (Bauman 2001) 'coalition proxy warfare' (Mumford 2013); 'transnational shadow wars' (Niva 2013); 'surrogate warfare' (Krieg and Rickli 2018), 'vicarious warfare' (Waldman 2018), 'liquid warfare' (Demmers and Gould 2018) or, simply 'remote warfare' (Biegon and Watts 2017). If we look beyond the labels, however, we notice how authors largely rely on three formats (or genres) to explain the shift to what is considered a new way of war: democratic risk-aversion, technology and networking.

Democratic risk-aversion

Authors grouped within this genre point at the appeal of remote technologies of warfare. They argue that democracies, in particular, turn to remote warfare as a way of risk-aversion. Simply put, decision makers in democracies fear losses among their own constituencies more than authoritarian leaders, because rising numbers of casualties will have adverse effects on public support and decrease their chance of re-election (Freedman 2006, 7). For one, remote technologies such as unmanned systems give human soldiers the best possible force protection: they are not exposed to the enemy at all. Grounded in classic liberal thought, and often referring to Immanuel Kant's notion of perpetual peace, this strand of thinking sees the 'no body bags' call of the electorate in liberal democratic societies as restraining politicians from engaging in high-risk warfare. In his famous treatise, Kant ([1795] 1957, 12–13) provided important insight into the risk aversion of democracies. He argued that when those who decide to wage war are obliged to fight and bear the costs:

> [...] they would be *very cautious* in commencing such a poor game, decreeing for themselves all the calamities of war. Among the latter would be: having to fight, having to pay the costs of war from their own resources, having painfully to repair the devastation war leaves behind, and, to fill up the measure of evils, load themselves with a heavy national debt that would embitter peace itself and can never be liquidated on account of constant wars in the future (emphasis added).

Remote warfare, in many ways, helps overcome this problem of 'costs', both in terms of human lives and expenditure. The end of conscription in Western democracies, with Vietnam as an important turning point, already formed a first step of transferring risks to military professionals. But with the emergence of remote technology wars can now be fought from a distance, allowing for zero-risk warfare. Soon after the invasion of Iraq, Martin Shaw (2005) in his book on the 'new western way of war' argued that liberal democracies aim to 'transfer the risks of war' even further: away from their own professional soldiers to the civilians and armed actors of 'the enemy.' In liberal democracies, warfare has become primarily an exercise in risk-management. For Shaw, this explains the strong preference for long-distance air strikes and drones, instead of military interventions with ground troops. In a similar vein, Coker (2009) and, more critically, Sauer and Schörnig (2012) refer to 'war in an age of risk', and 'democratic warfare' to highlight how democratic institutions and their publics are the central factors constituting the turn to remote warfare.

Technology

The role of technology, and particularly the turn to military robotics and autonomous weapons, figures prominently in the second explanation of remote warfare. Although the relationship between technology and war is an old one, the recent revolution in military technology and the emergence of a 'military-tech complex' is seen by some to be the main driving force behind remote warfare. Unmanned aerial vehicles (or drones), in particular, offer unprecedented possibilities to wage war from a distance. In addition to being seen to 'save lives' of both military personnel and civilians 'on the ground' and reduce costs, these systems offer numerous advantages to the military. As pointed out by Sauer and Schörnig (2012, 363): 'Machines can operate in hazardous environments; they require no minimum hygienic standards; they do not need training; and they can be sent from the factory straight to the frontline, sometimes even with the memory of a destroyed predecessor.' Within International Relations, a new subfield of drone studies has emerged with a strong emphasis on materiality, the politics of 'things', the agentic capacity of drones, the absence of human bodies, and the 'ethics of killing' (see: Holmqvits 2013; Schwarz 2016; Wall and Monahan 2011; Walters 2014; Wilcox 2017). Scholars working from the field of political economy have added to this second explanatory genre by pointing out how we have entered a new – and highly profitable – arms race, with tech corporations such as Microsoft, Amazon, Palantir and Anduril feeding the remote war machine. The basic viewpoint here is that artificial intelligence, autonomous systems, ubiquitous sensors, advanced manufacturing, and quantum science will radically push for – and transform – remote warfare.

Networking

A third, and equally prominent, genre refers to the key importance of the networked nature of contemporary warfare. Ever since the publication of Stanley McChrystal's 2011 article in *Foreign Policy,* the notion that 'it takes a network to defeat a network' has become somewhat of a standard, in both military and academic analysis (McChrystal 2011). Simply put, the argument goes that because the 'enemies of the state' are now operating through shadowy networks and cells, the state must resort to similar tactics. Reflecting upon this, Niva (2013) emphasises how US wars, legitimised by the War on Terror, have become increasingly 'networked', calling them 'transnational shadow wars.' While recognising the role of military robotics as an essential tool, he contends that the central transformation enabling remote warfare has less to do with new technologies and more to do with new forms of social organisation – namely, 'the increasing emergence of network forms of organisation within and across the US military after 2001' (Niva 2013, 187). In these new forms of 'counter-netwars' hybrid blends of hierarchies and networks, consisting of special operations forces, intelligence operatives, and private military contractors, have mounted strike operations across shadowy transnational battle zones (Ibid., 187). During the 2010s, this has mutated to heavily rely on what is called 'security cooperation.' Herein, small (private) military teams train, equip and advise local forces to do the fighting and dying on behalf of Western 'boots on the ground' (Biegon and Watts 2017). What is implied in this final genre is that remote warfare results from the state mimicking its enemies.

Perpetual warfare

Although the above debates each provide important insights, they need to be combined to create a comprehensive, multifactor explanation. Whereas we see democratic risk aversion as the key driver behind the Western turn to remote warfare, this new way of war also heavily depends upon technological advancements, political economies, and the outsourcing of the burden of war to private and distant others. What the above debates – even if combined – fail to address, however, is the more fundamental question on the transformative capacity of violence. That is, on how this new type of war is able to (re)work relations of power, order and politics. In thinking through the main moral and political challenges that remote warfare has given rise to, our concern with the proliferation of this type of violence lies in its normalisation of the asymmetrical distribution of death and suffering. Returning to Kant's famous words, remote warfare presents us with a paradox: if indeed remote technologies help to overcome democracies' casualty-sensitivity, and if 'bloodless war' (Mandel 2004) becomes a reality, will these democracies then not become less 'cautious' in commencing the 'poor game' of war? Or as Michael Ignatieff (2000, 4) phrased it in 2000 after the NATO bombings of Serbia and Kosovo – the first riskless war in history – 'if war becomes unreal to the citizens of modern democracies, will they care enough to restrain and control the violence exercised in their name?.' With zero direct risks and no returning body bags, we foresee that the perpetual peace doctrine will come to facilitate perpetual war.

Counting the bodies of the dead

Let us take a closer look at the manufacturing of the 'lack of care' that remote technologies facilitate: how have Western publics responded so far? On a positive note, we have witnessed the advent of a set of Western civil society organisations that monitor the local impact of remote warfare interventions and demand transparency and accountability for its harmful effects. Not satisfied with how Western militaries assess the number of civilians killed by their air strikes through relying on internal visual military intelligence recorded from the sky, organisations, such as Airwars, Amnesty International and Human Rights Watch, have developed new remote sensing techniques to count the number of civilian casualties from Western air strikes. They use open-source intelligence (such as social media posts and satellite imagery) to track, triangulate and geolocate, in real-time, local claims of civilian casualties. Concurrently, they monitor and archive official military reports on munition and strike statistics to measure them against the public record and grade the reliability of the claims made. In some cases, this is augmented by investigations on the ground.

The underlying assumption driving these initiatives is that publics do care, but simply need to be informed. If Western publics and parliaments are provided with systematic real-time evidence of the devastating effects of remote interventions for the civilians besieged, they will press their governments to constrain this way of war. An illustrative example is how Amnesty International and Airwars joined forces to monitor the impact of the 2017 US-led bombing campaign to retake the Islamic State-held city of Raqqa in Syria.

In the four-month remote Battle for Raqqa (June–October 2017), the US, UK and France fired over 40,000 air and artillery strikes that were called in by their local allies the Syrian Democratic Forces.[1] In the immediate aftermath of what was described by the US Defence Secretary James Mattis as a 'war of annihilation', the Coalition acknowledged just 23 civilian casualties, yet refused to conduct any investigations on the ground. In response, Amnesty International and Airwars set up an innovative crowdsourcing data project

---

[1]   These artillery strikes were fired from 50 km outside of Raqqa.

called *Strike Tracker*.[2] This online project engaged over 3000 digital activists from across 124 countries to help them trace and geolocate how the Coalition's bombings destroyed almost 80 percent of Raqqa. Drawing evidence from social media posts, two years of on-the-ground investigations, and expert military and geospatial analysis, they then built a database of more than 1,600 civilians reportedly killed in Coalition strikes, of which they were able to name 641. For the war as a whole, the US-led anti-ISIS Coalition engaged in over 34,000 strikes, firing over 100,000 munitions across Syria and Iraq from 2014 onwards. The resulting death toll is staggeringly different: Airwars estimates a minimum of between 8,214 and 13,125 civilian deaths, while the Coalition acknowledges 1,335.

Killing with care

This painstaking and often harrowing evidence-based monitoring work illustrates that some people in Western democracies obviously care. We, however, observe that representatives of the US-led Coalition also take great care to effectively thwart the critical counterclaims made by these watchdog and human rights organisations on the human suffering caused by their air strikes. We identify three ways in which this occurs. First, we see a lack of international media coverage of the numbers produced. Second, an outright denial by official sources. Third, and closely related, Coalition partners are quick to justify their own violence using classic war tropes.

In contrast to Western media's near-daily reporting and moral outcry over Russia's remote bombardments in Aleppo or Idlib, there was hardly any real-time coverage of the Coalition's bombing of Raqqa (see O'Brien 2019). Two years after the fact, most major news outlets (including CNN, BCC and Reuters) did pay lip service to the 1,600 body count. All reports, however, quoted the Coalition spokesperson's acknowledgement of just 159 of these allegations (thus denying the other 90 percent) and his justification:

> Any unintentional loss of life during the defeat of [IS] is tragic. However, it must be balanced against the risk of enabling [IS] to continue terrorist activities, causing pain and suffering to anyone they choose. The coalition methodically employs significant measures to minimise civilian casualties and always balances the risk of conducting a strike against the cost of not striking.[3]

---

[2]   See Strike Tracker https://decoders.amnesty.org/projects/strike-tracker

[3]   See BBC. 2019. 'IS conflict: Coalition strikes on Raqqa 'killed 1,600 civilians.' 25 April 2019. https://www.bbc.com/news/world-middle-east-48044115; Reuters. 2019. "Amnesty, monitors say US-led coalition killed 1,600 civilians in Syria's Raqqa." April

If we zoom out, we see here that remote wars, like all wars, do not remain insulated from the machinations of propaganda. War is a high-stakes enterprise, and public perceptions and public support are never left to chance (see Griffin 2010). Remote warfare is shrouded in denial and secrecy. Still, governments strive to justify their 'distant wars', while journalists often toe the party line to avoid being named unpatriotic or being 'blocked out' from powerful and official sources. It is these pressures and considerations which often result in a news diet of steady and highly standardised portrayals.

For the case of Operation Inherent Resolve, those in favour contrast the brutal and barbaric violence perpetrated by ISIS with the surgical precision with which its strongholds were targeted. This distinction between 'their' violence as vicious and barbaric and 'our' violence as clean and precise fits the classic tropes of war. Such statements suggest a good deal about how we like to understand our own violence. They establish a highly appealing contrast between borderland traits of barbarity, excess and irrationality, and metropolitan characteristics of civility, restraint and rationality (see Duffield 2002, 1052). In reaction to Airwars' high civilian body count, Coalition commander Stephen J. Townsend, for instance, claimed: 'Assertions by Airwars [...] and media outlets that cite them, are often unsupported by fact and serve only to strengthen the Islamic State's hold on civilians, placing civilians at greater risk.' Townsend emphasised that the Coalition dealt in facts and that he challenged anyone to find a more 'precise air campaign in the history of warfare ….The Coalition's goal is always for zero human casualties' (2018).[4] Or, as former US Defense Secretary James Mattis (2017) emphasised: 'We are the good guys… We do everything humanly possible consistent with military necessity, taking many chances to avoid civilian casualties at all costs'.[5] Such statements not only undermine the authority and legitimacy of the monitoring organisations, but they repeatedly point out how the constant application of new smart technologies and proportionality principles allow for a new form of perfect warfare, which saves lives of both Western military personnel and friendly civilians on the ground.

25, 2019, https://www.reuters.com/article/us-syria-security-raqqa/amnesty-monitors-say-u-s-led-coalition-killed-1600-civilians-in-syrias-raqqa-idUSKCN1S11HM; CNN. 2019. "Report: US-led coalition killed 1,600 civilians in Raqqa in 2017." April 25, 2019. https://edition.cnn.com/2019/04/25/politics/amnesty-international-report-raqqa/index.html.

4   See for transcript statement: Airwars. 2017. 'Former Coalition Commander Lt Gen Townsend responds to Airwars article on Raqqa.' 15 September 2017. https://airwars.org/news-and-investigations/former-coalition-commander-lt-gen-townsend-responds-to-airwars-article-on-raqqa/.

5   See for transcript interview: CBS News. 2017. 'Transcript: Defense Secretary James Mattis on "Face the Nation."'28 May 2017. https://www.cbsnews.com/news/transcript-defense-secretary-james-mattis-on-face-the-nation-may-28-2017/.

Conversely, critical voices point out the dangers of these forms of 'ethical killing' (Schwarz 2016) or 'humanitised violence' (Bonds 2018). They take issue with what they see as the production of a new type of ethical proposition that, paradoxically, presents 'killing as a moral act of care' (Chamayou 2015, 139). By framing acts of military violence in medical terms ('surgery', and 'precision'), we are made to imagine we are killing with care, and it then becomes hard to care for those who are killed carefully. This sanitisation directs our attention away from what is essentially a political act; coalition state violence needs to be accounted for, both legally and politically. Why was the operation launched in the first place? What was the international legal mandate to do so? And if indeed the bombings are legitimised as 'collective self-defense', is this how Western democracies best protect their own and local citizens against armed attacks? What are the boomerang effects of destroying 80 percent of a city with 'utmost precision'? More complicatedly, and perhaps painfully, why and how was ISIS able to emerge? How was the West involved in creating the conditions for ISIS' explosive success? We acknowledge that directly addressing particular wars and militarism is intensely political. These questions, therefore, require careful analysis, consideration and open debate. In this, we need to move beyond discourses of precision and sanitisation. For Chamayou, 'precision killings are still killings' (2015, 140). Or as Hannah Arendt argued much earlier: 'Politically, the weakness of the argument has always been that those who choose the lesser evil forget very quickly that they chose evil' ([1964] 2003, 36–37).

Conclusion

To be clear, we do not argue for more Western body bags or more boots on the ground. What we emphasise in this chapter is the need to 'make strange' the evolving normalisation of remote warfare as the lesser evil – as precise, efficient wars of necessity. Western democracies have largely removed their military from the theatres of war. As we saw in the opening vignettes, Western soldiers thereby no longer engage in 'hostilities' directly. But this does not make them any less violent.

Apart from 'making visible' the direct suffering caused by remote warfare, we aim to think through the transformative effects and moral and political challenges that this new way of war has given rise to. Key to the continuation of remote warfare, in addition to the secrecy of its operations and the sophisticated propaganda of precision and care, is its asymmetrical distribution of death and suffering. As we have seen above, zero-risk warfare is compelling to those not at the receiving end of the violence. Altogether, we conjecture this to translate into liberal democracies becoming not less, but more, war prone. That is the paradox of 'democratic warfare', and herein, we argue, lies its transformative effect. The violence is executed so remotely, that it becomes invisible, uncared for, and even ceases to be defined as such.

A second concern is blowback. The challenges that remote killings present to the logic of warfare also have serious political implications. As pointed out by Ignatieff (2000) and Sauer and Schörnig (2012), the riskless setup of remote warfare could very well justify a mirroring of 'remote' ways of fighting in the form of terrorist attacks by the enemy as the only possible way to retaliate in the absence of a 'fair' fight. As for the secrecy of remote violence, one thing is clear: in this age of digital media everything can be seen, filmed, and shared. Violence always has a boomerang effect.

In sum, outsourcing the violent act to robotic, private or surrogate others has silently taken political deliberation out of contemporary warfare. As a consequence, this has lowered the threshold for military engagement in liberal democracies. We need to bring politics, and the public, back in. Although this evidently entails a much wider and more profound discussion, we here conclude by making two suggestions. What we can detect from our opening vignettes is that, for a start, the new strategies for military engagement that come with remote warfare have to find a reflection in new political decision-making procedures. Any form of military intervention, whether offensive or defensive, that results in acts of physical harm on the ground should eventually be put through careful parliamentary scrutiny (such as, for the case of the US, the War Powers Resolution). This is what 'engaging in hostilities' should mean: inflicting harm to enemy combatants or civilians. Second, more analysis, dissemination and debate on the intimate realities of remote warfare is needed. Hopefully, this contribution has provided some useful insights into that direction.

Finally, Western democracies' claims to the moral high ground with respect to the brutality of war are uncalled for. All war is terrible, whether it is executed by a soldier piloting a weaponised drone or an insurgent's improvised explosive device. There is no such thing as sophisticated violence.

References

Airwars. 2017. 'Former Coalition Commander Lt Gen Townsend responds to Airwars article on Raqqa.' September 15.

Arendt, Hannah. [1964] 2003. *Responsibility and Judgment*. New York: Schocken Books.

Bauman, Zygmunt. 2001. 'Wars of the Globalization Era.' *European Journal for Social Theory,* 4(1): 11–28. https://doi.org/10.1177%2F13684310122224966.

BBC News. 2019. 'IS conflict: Coalition strikes on Raqqa "killed 1,600 civilians"'. April 25.

Biegon, Rubrick, and Tom Watts. 2017. 'Defining Remote Warfare: Security Cooperation.' Remote Control Project, Oxford Research Group. November. https://www.oxfordresearchgroup.org.uk/Handlers/Download.ashx?IDMF=0232e573-f6d6-455e-9d34-0436925002d4

Bonds, Eirk. 2019. 'Humanitized violence: Targeted killings and civilian deaths in the US war against the Islamic State.' *Current Sociology*, 67 (3): 438–455. https://doi.org/10.1177%2F0011392118807527.

CBS News. 2017. 'Transcript: Defense Secretary James Mattis on "Face the Nation."' May 28.

Chamayou, Grégoire. 2015. *Drone Theory*. London: Penguin Books.

CNN. 2019. 'Report: US-led coalition killed 1,600 civilians in Raqqa in 2017.' 25 April.

Coker, Christopher. 2009. *War in an age of risk.* Cambridge: Polity Press.

Dahhan, Ghassan and Milena Holdert. 2018, 'Hoe de Nederlandse overheid een Syrische 'terreurbeweging' faciliteerde. ' *Trouw*. 10 September. https://www.trouw.nl/nieuws/hoe-de-nederlandse-overheid-een-syrische-terreurbeweging-faciliteerde~bd73dd6e/

Demmers, Jolle and Lauren Gould. 2018. 'An Assemblage Approach to Liquid Warfare: AFRICOM and the 'hunt' for Joseph Kony.' *Security Dialogue,* 49(5): 364–381. https://doi.org/10.1177%2F0967010618777890.

Freedman, Lawrence. 2006. 'Iraq, liberal wars and illiberal containment.' *Survival,* 48(4): 51–65. https://doi.org/10.1080/00396330601062709.

Griffin, Michael. 2010. 'Media images of war.' *Media, War and Conflict,* 3(1) April: 7–41. https://journals.sagepub.com/doi/abs/10.1177/1750635210356813

Holmqvist, Caroline. 2013. 'Undoing war: War ontologies and the materiality of drone warfare.' *Millennium,* 41(3): 535–552.

Ignatieff, Michael. 2000. *Virtual War: Kosovo and Beyond.* London: Picador.

Kant, Immanuel. [1795] 1957. *Perpetual Peace*. New York: Bobbs-Merrill.

Krieg, Andreas, and Jean-Marc Rickli. 2018. 'Surrogate Warfare: The Art of War in the 21st Century.' *Defence Studies,* 18(2): 113–130. https://doi.org/10.1080/14702436.2018.1429218

Mandel, Robert. 2004. *Security, Strategy, and the Quest for Bloodless War*. Boulder, CO: Lunnen Rienner.

McChrystal, Stanley. 2011. 'It Takes a Network.' *Foreign Policy.* 12 February.

Mumford, Andrew. 2013. 'Proxy warfare and the Future of Conflict.' *RUSI Journal,* 158(2): 40–46. https://doi.org/10.1080/03071847.2013.787733

Niva, Steve. 2013. 'Disappearing violence: JSOC and the Pentagon's new cartography of networked warfare.' *Security Dialogue,* 44(3): 185–202. https://doi.org/10.1177%2F0967010613485869.

O'Brien, Alexa. 2019. 'News in Brief: US Media Coverage of Civilian Harm in the War Against So Called Islamic State.' Airwars. July. https://airwars.org/wp-content/uploads/2019/07/Airwars-News-in-Brief-US-media-reporting-of-civilian-harm.pdf

Olsen, Gunar. 2019. 'Add Trump's Yemen Veto to Obama's Spotty War Legacy.' *The New Republic,* April 20. https://newrepublic.com/article/153639/add-trumps-yemen-veto-obamas-spotty-war-legacy.

Reuters. 2019. 'Amnesty, monitors say US-led coalition killed 1,600 civilians in Syria's Raqqa.' 25 April.

Sauer, Frank and Niklas Schörnig. 2012. 'Killer drones: The "silver bullet" of democratic warfare?' *Security Dialogue,* 43(4) 363–380. https://doi.org/10.1177%F0967010612450207

Schwarz, Elke. 2016. 'Prescription drones: On the techno-biopolitical regimes of contemporary 'ethical killing'.' *Security Dialogue,* 47(1): 59–75. https://doi.org/10.1177%2F0967010615601388

Shaw, Martin. 2005. *The New Western Way of War.* Cambridge: Polity Press.

UK Parliament. 2018. 'Special Forces: Question for Ministry of Defence.' 16 April.

US Department of State. 2011. 'Libya and War Powers.' 28 June. https://2009-2017.state.gov/s/l/releases/remarks/167250.htm

Waldman, Thomas. 2017. 'Vicarious Warfare: The Counterproductive Consequences of Modern American Military Practice.' *Contemporary Security Studies,* 39(2): 181–205. https://doi.org/10.1080/13523260.2017.1393201

Wall, Tyler, and Monahan Torin. 2011. 'Surveillance and violence from afar: The politics of drones and liminal security-scapes.' *Theoretical Criminology,* 15(3): 239–254. https://doi.org/10.1177%2F1362480610396650

Walters, William. 2014. 'Drone strikes, *dingpolitik* and beyond: Furthering the debate on materiality and security.' *Security Dialogue,* 45(2): 101–118. https://doi.org/10.1177%2F0967010613519162

Wilcox, Lauren. 2017. 'Embodying algorithmic war: Gender, race, and the posthuman in drone warfare.' *Security Dialogue,* 48(1): 11–28.

3

Intelligence Sharing in Remote Warfare

JULIAN RICHARDS

In the post-9/11 period, the logic of remote warfare for Western powers has been greatly affected by the challenging and transnational nature of terrorist and criminal movements, and by a growing Western fatigue with fatalities amongst its own troops. Increasing budgetary pressures on military expenditure and the drive to 'achieve more with less' are undoubtedly increasing the lure. Coupled with these drivers, advancements in technology are encouraging Western nations to establish relationships and capabilities with partners that allow for intelligence collection from afar. These developments can offer security dividends if conducted effectively but can also come with a potential cost to state and society. This chapter examines the role that intelligence sharing plays in the broader concept of remote warfare and evaluates the likely risks to state and society. It considers the ways in which intelligence sharing underpins developments, in the shape of the sharing of bulk data at speed and the networking of weapons systems. In a sense, intelligence is the glue that binds together partners and agents in the whole development of the remote warfare landscape.

There are undoubtedly strong drivers to develop and enhance intelligence sharing relationships in the modern environment of conflict and risk (Aldrich 2004; Reveron 2006; Richards 2018), and these are evaluated here. Not all of these drivers are necessarily nefarious, and, if safeguards are observed, intelligence sharing has the potential to make the world a safer place. If done badly, however, the sharing of intelligence can run the risk of outsourcing legally and ethically dubious activities to those states who do not share the same standards of human rights and democratic accountability in their pursuit of national security (Krishnan 2011). In the case of a country such as the UK, the more partners with whom intelligence is shared and the worse their respective histories of human rights compliance, the greater the challenges faced in convincing others that security is being delivered in a democratic, accountable and ethical way. A case study is then examined of the UK in the post-9/11 environment, and the challenges it has faced in its intelligence sharing activities.

A related danger concerns the 'bulk' sharing of intercepted material, as Edward Snowden revealed was happening between the US and multiple allies, including the UK, in his release of classified material in 2013. Here, the risk is that highly complex and integrated signals intelligence (Sigint) systems sharing ever more industrial-scale amounts of data, could allow for unverified misuse of intelligence. There is a risk to privacy here as much as a risk of abuse.

Added to these problems is the fact that a state's oversight of its intelligence agencies and their activities can be inherently difficult (Phythian 2007; Gill 2012; Dobson 2019). Within this landscape, intelligence sharing relationships are often among the most sensitive aspects of any intelligence agency's operations. Such relationships are usually shrouded in heavy secrecy, not only from the public but occasionally from a state's own oversight bodies. States will argue national security reasons for this needing to be so, but going forwards, the importance of due diligence and robust oversight of intelligence sharing relationships and operations will need to be highly developed if serious risks to state and society are not to be realised.

The case for intelligence sharing

In many ways, the basic logic of intelligence sharing is difficult to dispute. Indeed, in response to the threat posed by violent extremists returning from conflicts such as those in Iraq and Syria (the 'foreign fighters' problem), the UN Security Council (UNSC) passed Resolution 2396 in 2017, reminding member states of the need for 'timely information sharing, through appropriate channels and arrangements' to disrupt the planning of attacks (UNSC 2017, 3).

As the erstwhile Director-General of Britain's MI5 intelligence agency, Eliza Manningham-Buller, noted (ISC 2018a, 134), the 9/11 attacks marked a watershed following which 'the need for enhanced international cooperation to combat the threat from al-Qaida and its affiliates' was taken as a given. Such threats from international terrorism have become more dynamic, with new connections and lines of information being forged across the globe with increasing ease and rapidity.

In the intelligence world, the 'Five Eyes' relationship which flowed from shared experiences in the Second World War, encompasses highly integrated intelligence sharing between the US, UK, Canada, Australia and New Zealand. Intelligence sharing operates on several other levels, however, many of which are far less structured and avowed than the Five Eyes or NATO. In some cases, a collection of states will participate in semi-structured, multilateral fora for sharing intelligence – a good example being the Club of Berne's group of Western security agencies (Walsh 2006), whose membership closely mirrors that of NATO.[1] At the tactical level, particular agencies will also sometimes participate in multinational intelligence 'hubs' or 'fusion centres'[2], usually dealing with specific issues such as regional counter-crime or counterterrorism. Beneath all of these more formal relationships, a myriad of bilateral or multilateral intelligence relationships will operate between states, with very focused objectives and mechanisms.

In all cases, intelligence sharing is a particularly sensitive and secretive business. The lifeblood of any security agency is the set of covert sources and capabilities it is able to deploy in ways that garner strategic advantage over adversaries (Warner 2002). The loss or compromise of such capabilities can lead to instant operational failure, and often political ignominy. Like reputations, sensitive intelligence sources take a long time to establish, but can be destroyed very quickly. Forging a relationship with a partner can often be about a complex web of mutual interests, whereby information is just one of the standards of currency.

Geography is usually crucial in prompting a relationship. In a sense, this is a key catalyst for remote warfare, as national security threats migrate out to the badlands of Asia, Africa and the Middle East. Such considerations provide the rationale for capacity-building projects, through which investments can be made in the capability of local partners. In the Five Eyes context, the dispersed geography of the partners was useful in establishing global interception systems such as ECHELON (Perrone 2001). More recently, evidence suggests that a number of airbases in Europe provide crucial communications infrastructure for directing the US' remote targeting across the Middle East, North Africa and South Asia (Amnesty International 2018, 6).

Such relationships may be asymmetric in the sense that the state reaching out to establish the partnership may receive benefits in a different area in return. These might not even be about intelligence capabilities *per se*, but

---

[1]   At the time of writing, the impact of Brexit on intelligence sharing relationships is unknown and subject to much conjecture (Inkster 2016; Hillebrand 2017).

[2]   Examples include Interpol, Europol, CARICOM's Regional Intelligence Fusion Centre (RIFC) in the Caribbean region, or the Central Asia Regional Information and Coordination Centre (CARICC), to name but a few.

could encompass military aid or other economic investments. This also means that such relationships can work both ways and that threats can be made to 'turn off the tap' if there are political or diplomatic problems – as Pakistan, for example, has frequently suggested to the US (Bokhari et al 2018). In many ways, this mirrors the wider problem of perverse incentives created by long-term military aid programmes, of which intelligence capacity-building is often a part (Bapat 2011; Boutton 2014).

There is a particular factor here concerning terrorism. One of the key benefits is that counterterrorism (like counter-crime) tends to transcend all other political considerations, even if definitions of who the 'terrorist' is can vary considerably in the face of local political objectives. That aside, from a policy perspective, the basic strategic concept of countering transnational terrorism can be the one topic on which virtually every state agrees, even if they do not in most other aspects. This applies to Western relationships with Russia and China, for example, and to relationships with Middle Eastern states.

Difficulties and challenges

A key principle of intelligence sharing is the 'third-party rule', which means that any country receiving intelligence from a partner agrees not to share it onwards with another party – unless they have express permission to do so. This agreement relies on mutual trust and it is not always possible to be certain where a piece of intelligence has ended up. There is, of course, also the constant risk that a partner agency may be infiltrated or corrupted by a hostile power.

A number of recent inquiries into intelligence activity have established that intelligence sharing relationships with international partners are rarely the subject of formal and documented memoranda of understanding (MoUs). Indeed, agencies such as MI6 point out that such formal arrangements are usually avoided, not only in order to keep the details to the minimum, but also because a fundamental lack of trust can be implied if the UK always insists on everything being formally documented and bureaucratised (ISC 2018b, 62). For an agency whose business is establishing relationships with states outside of the West with a different culture of bureaucratic norms, such factors must be taken carefully into account. On the other hand, as a former Ambassador to Uzbekistan noted, not documenting joint intelligence activities can sometimes turn out to be for reasons of the concealment of abusive behaviours (ISC 2018a, 60).

'Diplomatic assurances' are the formal method whereby intelligence partners commit to safeguarding human rights, and these have been established with several partner countries in the post-9/11 period. But human rights organisations such as Human Rights Watch (HRW) are scathing about the utility of such instruments as a safeguard against abuse (HRW 2005, 3). Amnesty International has echoed their sentiments, noting that 'the best way to prevent torture is to refuse to send people to places where they risk being harmed' (cited in Richards 2013, 183).[3]

It is the case that most non-Western states do not have clearly delineated and articulated expressions of their national security objectives and strategy (see for example HMG n.d.). In many cases, national security is just what a state must do to protect itself. Most do not have any legislation governing the scope or modus operandi of their intelligence and security agencies, and many have severely lacking or compromised mechanisms for parliamentary scrutiny of their activities.

The founder of the Muslim Brotherhood in Egypt, Hassan Al-Banna, was right in his prediction that entrenched states in the Middle East would always wish to repress populist Islamist movements (Mitchell 1993, 30). Western countries generally share this objective, and this drives much contemporary intelligence sharing. But the problem is that the underlying conception of national security may be different between states, and sometimes dangerously so. The problem can often manifest itself in the partner country wishing to obtain intelligence on expatriate dissident movements rather than on 'terrorists' per se, as a quid pro quo for supplying intelligence on terrorist suspects. For the UK, where London has been lambasted in the past as a haven for radicals and dissidents (Foley 2013, 248), this can be an attractive element for countries that wish to obtain intelligence on London-based political oppositionists. Rudner (2004, 214) describes how Egypt and Jordan have both complained to the UK about its failure to supply them with intelligence on dissidents residing in London, while Sepper (2010, 175) describes the case of the Libyan authorities being able to interrogate detainees at Guantanamo Bay about dissidents in the UK.

Conversely, intelligence provided to such countries on purported terrorist targets can lead to violent actions being taken on the ground, violating human rights, neutralising potential further sources of intelligence, and generating political blowback. After 1981, the US allegedly slowed the flow of intelligence to Mossad after the Israelis had used their information to destroy Iraq's nascent nuclear reactor in a pre-emptive military strike (Kahana 2001, 414). More recently, heavy military actions against Hamas and Hezbollah within the

---

[3]   Amnesty International, *'Europe must halt unreliable 'diplomatic assurances' that risk torture.* http://www.amnesty.org/en/news-and-updates/report/europe-must-halt-unreliable-diplomatic-assurances-risk-torture-2010-04-12

Occupied Territories continue to place Western military and intelligence partners of Israel in uncomfortable positions concerning complicity with disproportionate military action in civilian areas (Curtis 2018).

In many situations, war and violent counter-insurgency operations may cause especially difficult questions to be asked, not just in terms of the use of military equipment being supplied to repressive regimes, but also to the tactical use of intelligence. In the ongoing civil war in Yemen, for example, the US has come under increasing pressure to curb military and intelligence support to Saudi Arabia following destructive bombings that have caused considerable civilian casualties (Gambino 2018), not to mention a humanitarian catastrophe affecting much of the population. Britain's MI6 and Special Forces have also been implicated in supplying geolocational intelligence to the Americans to facilitate drone strikes by forces in the region (Norton-Taylor 2016). Such operations are framed by the states in question as tackling 'upstream' terrorist threats from the likes of al-Qaeda in the Arabian Peninsula (AQAP). But the question has to be asked – to what cost?

Case study: the UK's post 9/11 security environment

Officially, the UK makes a great deal of its mission to uphold values in its foreign policy. On the occasion of the 2017 International Day in Support of Victims of Torture, the Foreign and Commonwealth Office's (FCO) Minister for Human Rights, Lord Ahmad, noted that 'The UK government condemns torture in all circumstances' (FCO 2017). Urging other states to 'sign, ratify and implement' the UN Convention Against Torture and its Optional Protocol can feel disingenuous, however, when the UK itself becomes embroiled in detainee mistreatment scandals or arms sales to repressive regimes.

In Afghanistan in the post-9/11 period, operational collaboration with the new intelligence agency, the National Directorate of Security (NDS), has proved to be a complicated business. In 2007, Amnesty International revealed a catalogue of human rights abuses in Afghanistan and ISAF's alleged complicity in the abuse, much of it centred around the NDS's notorious 'Department 17' facility in Kabul (Richards 2013, 177–8). In 2012, the British peace activist Maya Evans was successful in securing a judicial review that placed a temporary moratorium on detainee handovers in Afghanistan (Carey 2013).

One of the more significant individual cases in the post-9/11 period was that of Binyam Mohamed, an Ethiopian national who had formerly been a resident in the UK. In April 2002, Mohamed alleges that he was arrested in Pakistan on terrorist charges and subsequently mistreated over a period of three months (ISC 2018a, 123–4). He alleged he was then illegally rendered to Morocco and thereafter to Guantanamo Bay, where he was subjected to further mistreatment (ISC 2018a, 124). In 2010, the UK Government announced that it had settled out of court with Mohamed and fifteen other former Guantanamo detainees, twelve of whom had launched legal action against the heads of MI5 and MI6, for undisclosed sums believed to number in the tens of millions of pounds (BBC News 2010).

The case of a Libyan dissident opposed to Muammar Gaddafi by the name of Abdel Hakim Belhaj caused similar political controversy. Belhaj was illegally rendered from Thailand to Libya by the CIA in 2004, acting on British intelligence (Hutton 2018). Allegations of subsequent brutal torture by the Libyans culminated in a claim against the British government for £1 in compensation and a full apology, eventually settled in May 2018, when a statement was delivered to parliament on behalf of the Prime Minister, apologising 'unreservedly' and lamenting Belhaj's 'appalling treatment' (Hutton 2018).

In both cases, the defining features were a willingness by UK intelligence agencies to work with unsatisfactory regimes to pursue their counter-terrorism objectives; and complicity in the mistreatment of detainees through a desire not to disrupt the key intelligence relationship with the US.

Meanwhile, one of the perpetrators of the 2013 murder of Fusilier Lee Rigby, Michael Adebolajo, has alleged that he was beaten and threatened with electrocution and rape on more than one occasion during detention in Kenya at the hands of a police unit with a relationship with British intelligence (ISC 2014, 153). Leaving aside his subsequent conviction for murder, the allegations highlighted a number of difficult questions for the British intelligence machinery on whether and how such allegations involving a partner country are investigated, and whether the UK is effectively complicit in mistreatment if one of its intelligence partners commits the wrongdoing. One major area of risk highlighted by the case was the question of which intelligence has been potentially derived from torture where multiple agencies were working together, and where intelligence is pooled in such a way that the provenance of individual pieces of information may be difficult to ascertain. The Chair of the Intelligence and Security Committee (ISC) has identified this as a significant area of ongoing risk.[4]

One of the more noteworthy investigations undertaken by the ISC in recent years has been that into the question of the mistreatment and rendition of detainees in the post-9/11 years (the Detainee Mistreatment and Rendition

---

4   Interview with author, 16 July 2018.

[DMR] Inquiry). This investigation struck at the heart of intelligence relationships with the UK's range of partners in the counterterrorism realm, with many of whom serious questions concerning human rights abuses were hanging in the air.

The problems in the early period after 9/11 were manifold. In all, the Inquiry found two cases where British intelligence officers appeared to have been directly involved in the mistreatment of detainees. There were 13 other cases where mistreatment was witnessed by British intelligence officers, and 128 cases where foreign intelligence partners spoke about the mistreatment of detainees. There were 232 documented cases where intelligence was shared with partners known to regularly practice mistreatment; and 198 cases where intelligence was received from such partners. Two instances were found of British intelligence agencies offering to pay for the extraordinary rendition of suspects; and 22 cases where British intelligence directly led to the illegal rendition of suspects.

The details amount to a comprehensive realisation during this period of the risk that intelligence relationships can lead to the serious compromise of human rights. Aside from some cases of apparent direct complicity in mistreatment, there was clear evidence of a lack of training amongst intelligence officers about what does or does not constitute mistreatment (ISC 2018a, 131). There was also evidence that parts of the British intelligence machinery either had no mechanism for filtering out intelligence that may have been derived from torture, or were generally happy to rely on broad assurances that standards were being upheld, when they should have had strong grounds for suspecting otherwise (Ibid., 55). On the key intelligence relationship with the Americans, the DMR Inquiry found evidence that British intelligence officers on the ground were either unwilling to raise questions about apparent mistreatment, or did so only half-heartedly, for fear that they would damage the overall intelligence relationship (Ibid., 58). This constituted a serious structural risk in the system.

From 2004 onwards, the DMR Inquiry found evidence of the situation starting to change for the better. In 2010, the Consolidated Guidance (CG) on how to properly handle detainees was issued to all intelligence officers on the ground. Sir Mark Waller, the Intelligence Services Commissioner for the period 2011–16, subsequently told the ISC that he was 'broadly happy' that the various intelligence services were selecting the right cases to which the CG should apply and were properly flagging up the cases in which there could be concerns (Ibid., 22).

The CG should not be viewed as a panacea, however. The ISC, and Sir Mark

Waller, have flagged a specific concern that the CG does not adequately address the broader context of intelligence relationships with joint units, but only case-specific incidents and exchanges (ISC 2018b, 50). The question is partly one of resources and capabilities, since perpetual monitoring of day-to-day conduct in an overseas joint unit is difficult, resource-intensive, and could be perceived as indicative of a fundamental lack of trust in the partner.

In some respects, this relates to the wider question of the utility and risks of capacity-building programmes in the modern era. As Watling and Shabibi (2018) noted in the context of Yemen, such programmes involving multiple partners can be complex, politically fraught, cost-intensive and difficult to bring to a stage where they are adding value on the ground rather than exacerbating existing problems and tensions. This is not to say that they are always redundant, however: the right programme, properly managed, can deliver important dividends.

Risks to state and society

The discussion thus far has highlighted the potential dilemma for modern states engaged in remote warfare to balance the imperatives of sharing intelligence with partners to deliver national security, against the risk of 'dirty hands' (Walzer 1973, 161) that arises in doing so. The principal risk is that increased flows of intelligence between partners may mean safeguarding human rights not only becomes more difficult to ensure, but that even knowing where rights have been compromised will be increasingly difficult to establish.

For liberal democratic states such as the UK, the first and most obvious risk is a reputational one, whereby supposed commitments to universal human rights can start to sound like empty promises when cases of complicity in abuse arise. This could, in turn, reduce the influence of the UK on the world stage at a time when it can ill afford to do so.

For broader society, there are fundamental questions about a retrenchment from the core values of peace, democracy and human rights. In the intelligence sharing context, there are also public fears about an inexorable creep towards a global 'surveillance society' (Beck 2002; Kerr and Earle 2013; Lyon 2014; Richards 2016). At a time when authoritarian regimes are increasingly managing to place national security imperatives above commitments to modern liberal values, states such as the UK should be aiming to be the vanguard of such liberal values, rather than allowing themselves to fall into the same boat of authoritarianism, secrecy and abuse.

The advent of 'Big Data' (which means both a massively increased amount of available data on the activities of the citizenry; but also increasingly sophisticated technology for extracting value from such data) has delivered a complex set of opportunities and risks for the major intelligence services. On the partnerships front, improving technology has increasingly allowed for industrial-scale pooling and cross-referring of major data collections spanning global communications, by linking-together the Sigint systems of partners. A secret National Security Agency (NSA) system uncovered by Snowden called RAMPART-A, for example, appears to be an international network of interception capabilities against trunk fibre-optic cables carrying the bulk of the global communications network (Gallagher 2014). The system is part of a network of 33 third-party Sigint relationships (Gallagher 2014).

Again, reputational issues concerning the conduct of liberal democratic states as opposed to those of authoritarian regimes such as China – who make no secret of the need to undertake near-ubiquitous surveillance of their citizenry – are placed on the table by such revelations.

As the civil rights NGO Privacy International (2018, 10) noted, there are three potential problems with these bulk surveillance activities. First is the question of the basic, extra-territorial human right to privacy. A related question is that of ensuring the legal protection against surveillance of the communications of a state's own nationals, and that of particularly sensitive interest-groups such as lawyers, doctors and journalists. Germany is one of the few countries that has taken steps to try to address this particular issue legislatively following a parliamentary inquiry[5], although in the view of one commentator, subsequent changes serve only to make oversight of the national intelligence service, the BND (Bundesnachrichtendienst*)*, even more confusing and fragmented (Wetzling 2017). In the UK, MI5 has recently been castigated for having 'lost control' of its data retention and handling in such a way that unlawful invasions of privacy may have become a systemic issue (Bowcott 2009).

In a case brought to the Investigatory Powers Tribunal by Privacy International against GCHQ in 2013 about access to an NSA system called PRISM (Privacy International 2018, 24), the parliamentary ISC committee found no evidence that GCHQ had been circumventing UK law through its access to the NSA system (ISC 2013). But, as with the abovementioned case against MI5's data handling, there may be a tendency amongst national intelligence services to conceal from their oversight bodies information that has not been explicitly requested. This could be either because something

---

[5]   *Die Gesetzes zur Ausland-Ausland Fernmeldeaufklärung des Bundenachtrichtendienstes*; Laws on Foreign-to-Foreign Intelliegnce Gathering of the Federal Intelligence Service.

serious is amiss, or simply because adequate procedures have not been followed properly. Such cases undermine trust in the integrity of the agencies and in the capabilities of the state's oversight function.

Amnesty International (2018) has outlined a set of concerns about intelligence sharing arrangements between a set of European countries and the CIA in the facilitation of lethal drone strikes through the provision of geolocational data. Given the number of non-combatant collateral casualties in such strikes, there is an ongoing debate as to whether such activities are legal under international law. In the Netherlands, the revelation of the scale and complexity of data exchanges with the US on Somali piracy has triggered a comprehensive inquiry by the state's parliamentary oversight body, the CTIVD (Commissie van Toezicht op de Inlichtingen). Indeed, legal challenges concerning intelligence assistance to the US in facilitating lethal drone strikes have been launched in several of the US's European intelligence partner countries (Amnesty International 2018, 7).

The fundamental question here is perhaps a deep-rooted and significant one about the impact of new technology on society. As with the advent of artificial intelligence (AI) and automation, one can foresee both exciting new opportunities and grave risks, depending on one's point of view. For intelligence services, galloping technology in the areas of data collection, mining and analysis, offer tremendous new opportunities for tackling complex international threat actors and delivering national security. But there are also manifold risks in sliding towards authoritarianism and repression, and many of these are only just beginning to take shape.

Going forwards

The de-centred and borderless nature of contemporary threats such as those posed by al-Qaeda or Islamic State, means there is an increasingly inescapable logic in sharing intelligence with as many cooperative partners across boundaries as possible. Again, technological developments in database capacities, bulk data transmission and algorithmic analysis have encouraged and enabled such transformations.

The UK discovered to its cost after 9/11, however, that some intelligence relationships can, in the wrong circumstances, lead to complicity in serious human rights abuses. In many cases, these arose from the importance of the relationship with the US and the perceived need not to damage that relationship. But alliances with other partners across the world who see national security in very different ways to us can also lead to problems. As the volumes of data shared and the automation by which such sharing happens both scale up, the ability to track back from a specific piece of information to the delivery of a human rights abuse becomes ever more difficult to achieve. There are serious moral questions to be asked about allowing such concerns to drift, especially in supposedly liberal democratic states.

Placing all of this in perspective, the answer is probably not to bolt the stable door completely. The fundamental drivers for sharing intelligence across boundaries in the pursuit of organised crime and terrorism are inescapable and are indeed mandated by the UN to all responsible member states. As with so many areas of society, new technologies can deliver tremendous benefits in this area if they are used responsibly.

The UK and partner states need to learn from the mistakes of the immediate post-9/11 period and ensure as much oversight and accountability of their intelligence sharing relationships as they can deliver. It is recognised, of course, that sensitive technologies and relationships should not be trumpeted on the front pages of the newspapers, since that will just help the enemies of democratic society. At the same time, liberal democratic societies need to ensure that in all areas of the move towards remote warfare, the importance of protecting rights and ensuring accountability will remain paramount.

Training and capacity-building of partners are not bad things and can indeed ensure that a rules-based and professional approach to security and intelligence becomes more widespread across states and society. Training and guidance for frontline officers working with partners also needs to be continually reviewed and developed.

In the rapidly developing area of data-sharing with partners, technology needs to ensure due diligence and audit functions for individual pieces of information as much as possible. To be fair, there is evidence that fears of outsourcing of illegal or unacceptable practices in this area have not been realised to any major extent, as far as can be determined. But the risks are rising continually as we move through the next major revolution in military affairs, and vigilance against eroding human rights needs to keep pace.

References

Aldrich, Richard J., 2004. 'Transatlantic Intelligence and Security Cooperation.' *International Affairs,* 80(4) (July): 731–53.

Amnesty International. 2018. *Deadly Assistance: The role of European states in US drone strikes.* London: Amnesty International.

BBC News. 2010. 'Compensation to Guantanamo detainees "was necessary".' 16 November. https://www.bbc.co.uk/news/uk-11769509

Beck, Ulrich. 2002. 'The Terrorist Threat: World Risk Society Revisited.' *Theory, Culture and Society,* 19(4): 39–55.

Bokhari, Fahran, Katrina Manson and Kiran Stacey. 2018. 'Pakistan halts intelligence sharing with US after aid suspension.' *Financial Times*. 11 January. https://www.ft.com/content/59969778-f6b1-11e7-88f7-5465a6ce1a00

Bowcott, Owen. 2019. 'MI5 accused of "extraordinary and persistent illegality."' *The Guardian*. 11 June. https://www.theguardian.com/uk-news/2019/jun/11/mi5-in-court-accused-of-extraordinary-and-persistent-illegality

Carey, Daniel. 2013. 'Maya Evans case: secret courts, torture and avoiding embarrassment.' *The Guardian*. 11 January. https://www.theguardian.com/law/2013/jan/11/maya-evans-secret-courts-torture

Curtis, Mark. 2018. 'The raw truth about the UK's special relationship with Israel.' *Middle East Eye*. 5 June. http://www.middleeasteye.net/columns/raw-truth-about-uk-israel-special-relations-456740882

Dobson, Melina J. 2019. 'The last forum of accountability? State secrecy, intelligence and freedom of information in the United Kingdom.' *The British Journal of Politics and International Relations,* 21(2: 3): 12–29.

FCO (Foreign and Commonwealth Office). 2017. 'UK government reaffirms its commitment to combat torture.' 26 June. https://www.gov.uk/government/news/uk-government-reaffirms-its-commitment-to-combat-torture

Foley, Frank. 2013. *Countering Terrorism in Britain and France: Institutions, Norms and the Shadow of the Past*. Cambridge: Cambridge University Press

Gallagher, Ryan. 2014. 'How secret partners expand NSA's surveillance dragnet.' *The Intercept*. 19 June. https://theintercept.com/2014/06/18/nsa-surveillance-secret-cable-partners-revealed-rampart-a/

Gambino, Lauren. 2018. 'Yemen war: senators push to end US support of Saudi Arabia.' *The Guardian*. 28 February. https://www.theguardian.com/world/2018/feb/28/yemen-saudi-arabia-war-us-support-senator-push-to-end

Gill, Peter. 2012. 'Intelligence, Threat, Risk, and the Challenge of Oversight.' *Intelligence and National Security,* 27(2): 206–22.

Hillebrand, Claudia. 2017. 'With or without you? The UK and information and intelligence sharing in the EU.' *Journal of Intelligence History,* 16(2): 91–94.

HMG. N.d. 'Fact Sheet 1: Our Approach to the National Security Strategy.' https://assets.publishing.service.gov.uk/government/uploads/system/uploads/attachment_data/file/62483/Factsheet1-Our-Approach-National-Security-Strategy.pdf

HRW (Human Rights Watch). 2005. '*Still at risk: Diplomatic Assurances no Safeguard against Torture.'* 17/4(D). April.

Hutton, Will. 2018. 'In the Belhaj case, Britain set aside the rule of law and moral principles.' *The Guardian.* 13 May. https://www.theguardian.com/commentisfree/2018/may/13/in-case-of-belhaj-britain-set-aside-rule-of-law-and-moral-principles

Inkster, Nigel. 2016. 'Brexit, Intelligence and Terrorism.' *Survival,* 58(3): 23–30.

ISC (Intelligence and Security Committee). 2013. *Press statement.* https://b1cba9b3-a-5e6631fd-s-sites.googlegroups.com/a/independent.gov.uk/isc/files/20130717_ISC_statement_GCHQ.pdf

———. 2014. *Report on the intelligence relating to the murder of Fusilier Lee Rigby.* London: TSO, HC 795.

———. 2018a. *Detainee Mistreatment and Rendition Inquiry 2001-10.* London: TSO, HC113.

———. 2018b. *Detainee Mistreatment and Rendition: Current Issues.* London: TSO, HC 1114.

Kahana, Ephraim. 2001. 'Mossad-CIA Cooperation.' *International Journal of Intelligence and Counterintelligence,* 14(3): 409–20.

Kerr, Ian, and Jessica Earle. 2013 'Prediction, Preemption, Presumption: How Big Data Threatens Big Picture Privacy.' *Stanford Law Review Online,* 66/65.

Krishnan, Armin. 2011. 'The Future of US Intelligence Outsourcing.' *Brown Journal of World Affairs,* 18(1): 195–211.

Lyon, David. 2014. 'Surveillance, Snowden, and Big Data: Capacities, consequences, critique.' *Big Data and Society,* July–December.

Mitchell, Richard P., 1993. T*he Society of Muslim Brothers.* New York: Oxford University Press.

Norton-Taylor, Richard. 2016. 'UK special forces and MI6 involved in Yemen bombing, report reveals.' *The Guardian*. 11 April. https://www.theguardian.com/news/defence-and-security-blog/2016/apr/11/uk-special-forces-and-mi6-involved-in-yemen-bombing-report-reveals

Perrone, Jane. 2001. 'The Echelon spy network.' *The Guardian*. 29 May. https://www.theguardian.com/world/2001/may/29/qanda.janeperrone

Phythian, Mark. 2007. 'The British experience with intelligence accountability.' *Intelligence and National Security,* 22(1): 75–99.

Privacy International. 2018. *Secret Global Surveillance Networks: Intelligence Sharing between Governments and the Need for Safeguards*. London: Privacy International. April.

Reveron, Derek S. 2006. 'Old Allies, New Friends: Intelligence sharing in the War on Terror.' *Orbis* (Summer 2006): 453–68.

Richards, Julian. 2013. 'Intelligence, Count-Insurgency and Reconstruction: Intelligence and International Cooperation in Afghanistan.' *Inteligencia y seguridad,* 13: 167–92.

Richards, Julian, 2016, 'Needles in Haystacks: Law, Capability, Ethics and Proportionality.' In *Big-Data Intelligence-Gathering,'* edited by Anno Bunnik, Anthony Cawley, Michael Mulqueen, and Andrej Zwitter. Basingstoke, Palgrave Macmillan.

Richards, Julian. 2018. *Defining Remote Warfare: Intelligence sharing after 9/11*. Remote Warfare Programme, Oxford Research Group.

Rudner, Martin. 2004. 'Hunters and Gatherers: The Intelligence Coalition Against Islamic Terrorism.' *International Journal of Intelligence and Counterintelligence,* 17(2) 193–230.

Sepper, Elizabeth. 2010. 'Democracy, Human Rights, and Intelligence Sharing.' *Texas International Law Journal,* 46: 151–207.

UNSC (Security Council). 2017. 'Resolution 2396.' 21 December. https://digitallibrary.un.org/record/1327675/files/S_RES_2396%282017%29-EN.pdf

Walsh, James. I. 2006. 'Intelligence sharing in the European Union: Institutions are not Enough.' *Journal of Common Market Studies,* 44(3): 625–43.

Walzer, Michael. 1973. 'Political Action: The Problem of Dirty Hands.' *Philosophy and Public Affairs,* 2 (2) Winter: 160–80.

Warner. Michael. 2002. 'Wanted: A Definition of "Intelligence". Washington DC: Center for the Study of Intelligence. January: 15–22.

Watling, Jack. and Namir Shabibi. 2018. 'British Training and Assistance Programmes in Yemen 2004 –2015.' Remote Warfare Programme, Oxford Research Group. June.

Wetzling, Thorsten. 2017. 'Germany's intelligence reform: More surveillance, modest restraints and inefficient controls.' *Stiftung Neue Verantwortung,* Policy Brief. June.

4

Remote Warfare and the Utility of Military and Security Contractors

CHRISTOPHER KINSEY AND HELENE OLSEN

This chapter explores the role of military and security[1] contractors in remote warfare. As the chapter explains, contractors in general have been a feature of warfare for centuries. This is not a coincidence, but rather the result of how warfare has been organised since the beginning of the sixteenth century, if not earlier. It was only after the rise of nationalism and the emergence of the modern state after the French Revolution, alongside industrialisation, and improvements in methods of bureaucracy that military contractors were temporarily marginalised, with many of their roles taken over by nationalised armed forces. This way of organising state violence survived until after the Second World War, when military and security contractors once more began to appear on the battlefield, supporting military operations, or conducting foreign military training. Today, they are a regular part of modern warfare.

This chapter is divided into five sections. The first section defines the key terms of remote warfare and military and security contractors. It is important to maintain clarity regarding the meaning of these terms as each has a range of contested meanings. The second section evaluates the practice of using military and security contractors in the contemporary operating environment. Section three seeks to explain the different rationales behind the phenomenon of military and security contracting and explores the practical challenges to the use of military and security contractors. Here, it would be easy to simply espouse the most obvious reasons why states rely on military and security contractors, namely as a means to reduce the political risk of

---

[1]   For the purpose of this chapter, the term 'military and security' is used because it provides a wider understanding of security.

deploying large numbers of military personnel and as a means to more efficiently manage military spending. However, this would provide an incomplete picture of military and security contracting. In this section, it will furthermore be argued that the challenges surrounding the use of military and security contractors are not going to disappear, but will only get harder to resolve over time if not confronted immediately or in the very near future. The fourth section will offer new insights into this important subject, arguing that the use of military and security contractors in remote warfare offers distinct opportunities for states, but only if the contractors are used in a mindful manner and with regard to key political, legal and ethical considerations. Importantly, it will suggest new ways for states to move forward when using military and security contractors to engage in remote warfare. Section five will offer concluding remarks.

Defining key terms

This section seeks to define the two concepts that are central to our argument: remote warfare and military and security contractors. For the purpose of this chapter, we will adopt a similar definition of remote warfare as Knowles and Watson (2018, 2–3). Drawing on their 2018 publication *Remote Warfare: Lessons Learnt from Contemporary Theatres*, which describes remote warfare in its most basic form, our definition describes the phenomenon as an approach to warfare that seeks to avoid the deployment of large numbers of ground forces in a military intervention, preferring instead to utilise an assortment of different activities and actors for a range of different purposes, on different timescales, and using different strategies. Continuing, such strategies may involve using special forces (SF), proxy forces, drones, and military contractors, separately or in combination with one another, to defeat an enemy. From a practical perspective, such an approach often involves coordinating a range of actors with various skill sets that are designed for different functions and with different operational and logistical requirements. Importantly, remote warfare is about countering an adversary. It must therefore be part of a military operation that is actively seeking out a foe to destroy. It may also be part of a wider array of military and security activities, such as supporting peacekeeping operations, that are being conducted alongside it (Østensen 2013, 33). Thus, in relying on the definition above it is easy to see why military and security contractors can be considered a component part of remote warfare.

What do we mean by military and security contractors and whom are we talking about? Starting with the second part of the question, military and security contractors are largely ex-military staff (Dunigan et al. 2013, xiv). This is the case for most contractors working for military and security companies that provide support services. Indeed, companies that provide these services often prefer employing former military staff because not only are they familiar with how the military operates, but they also find it easy to transition into a military environment, while at the same time feeling comfortable working in a conflict zone (Hawks 2014, chapter 5). The age of these individuals usually ranges from the mid-twenties to the mid-sixties (Ibid.), and the industry does not restrict its recruitment to a specific region but recruits across the globe (Dunigan et al. 2013, 9). However, there are a number of countries and regions that provide a significantly higher number of military and security contractors than others. One such country is Nepal. Many military and security companies prefer to hire retired Gurkhas because of their work ethic and loyalty. Management positions are usually filled by ex-military staff from Western countries, while the labour force is mainly comprised of ex-military staff from developing countries (Kinsey 2006b).

Military and security contractors are often viewed as mercenaries or soldiers of fortune – a view that puts emphasis on the actor's motivation and intention for participating in warfare as well as their active, front-line participation. To see them in this way, though, is simplistic. It fails to take account of the purposes behind their utility, where many military and security contractors operate far from the frontline, as well as the wide range of activities they perform. For example, today most military and security contractors perform vital, but non-combat roles, while combat roles remain the hallmark of the mercenary. However, the difference between mercenaries and military contractors is often found in the eyes of the beholder. Mercenaries have generally only been engaged during a short period of time on an *ad hoc* basis, while military and security contractors operate in more permanent structures during both peace and war. Abrahamsen and Williams (2011, 23) explain that there exists a 'mercenary misconception' when assessing military and security contractors that stems from the tendency to minimise the primarily non-military activities performed by these actors. Likewise, others have suggested that the character of the activities performed by these actors (moving in a spectrum between lethal/non-lethal) should determine the labelling of these individuals and companies as either mercenaries or military and security contractors (Kinsey 2006, 10). Similarly, Camm (2012, 144–145) groups military contractors under three headings that capture the extensive array of activities they now perform for deployed state military forces. These headings are *troop support* services, *system support* services in theatre, and *security protection* services (Ibid.). These categories can also apply to remote warfare, where military contractors provide services in support of a particular strategy directed at an adversary.

What, then, are included in these categories? *Troop support* services involve providing a broad range of services to support personnel. Examples of such services include catering, maintenance of infrastructure, waste disposal services and maintaining vehicles. Moreover, it is easy to imagine military and security contractors supporting a remote warfare strategy by providing these services to, for example, SF fighting a foe in a remote part of Africa.

The second category is *system support* services. These services include maintaining weapons and information technology (IT) systems. It is here where military and security contractors might operate as part of a remote warfare strategy outside the military chain of command, while still part of a country's security strategy, deploying or operating technology against enemy forces on behalf of another government agency. The third category is *security protection* services. These services usually involve armed security, sometimes heavily armed, but can also include police and military training. Most armed security roles involve protecting government facilities or government personnel abroad; for example, the guarding of embassies and the protection of high value individuals such as ambassadors. Again, the provision of armed security is more likely to be part of a broader approach, working with other actors, perhaps providing protection for their facilities, and not as a standalone security force. Where military and security contractors can act alone is in the military training of proxy forces as an alternative to using SF.

The use of private military and security in the contemporary operating environment

In many respects, the military and security contractor, or mercenary as some people prefer to call him or her, is the original remote warfare instrument, and the rationale behind employing him has changed very little over time. While the overall number of military contracts has fallen since the end of combat operations in Afghanistan and Iraq, the latter seeing as many as 100,000 contractors at the height of the conflict (Phelps 2016, 15), they are still very much a part of the modern battlefield. As Moore (2017, 20) points out in the case of the US military, contractors continue to be called upon to support military operations across the globe. Such support is varied, ranging from support for counter-terrorist operations around the Horn of Africa, counter-narcotics operations in Latin America, to working with military personnel at Al Udeid airbase in Qatar. The trend is the same with other Western state militaries, whilst Russia has also begun to rely more on military contractors than in the past. The Wagner Group has been operating in Syria since 2014, where it trains and fights alongside pro-government militias, while more recently it has undertaken operations in Sudan, [2] the Democratic Republic of

---

[2]   See https://www.theatlantic.com/international/archive/2018/08/russian-mercenaries-wagner-africa/568435/ (accessed 13 June 2019).

Congo,[3] Venezuela[4] and Libya[5]. This trend of utilising contractors does not seem to be slowing down, evidenced by other countries following the example of the US, UK and Russia.

Using military contractors is still seen as one way for states to overcome the practical constraints of undertaking expeditionary operations, as it allows states to deploy a light military footprint often to assist local forces through military training and advising. At the same time, it reduces the political risk of deploying national forces and helps to overcome the financial obstacles frequently attached to such operations. Indeed, military and security contractors are often seen as a cheaper alternative to deploying the state military, as they do not incur the same level of costs attached to state military personnel (for example, long-term medical care in the event of state soldiers being injured). Nonetheless, such an approach to warfare carries considerable risks as well as advantages. Nor is it easy to implement such an approach. At the moment, most countries employ military contractors on an *ad hoc* basis and not as part of a wider military strategy. Even so, this is an improvement from the Cold War, when contractors provided military training and security services to Third World armies and operated in the shadows of their home state's foreign policy. In past years, they often acted as the invisible hand of unofficial diplomacy, whereas now they operate openly with a legitimate business presence in many parts of the world. Moreover, like all major corporations, they no longer appear prepared to sacrifice their business interests for the interests of their home state.

So far, evidence on the performance of military contractors is mixed. Sierra Leone and Executive Outcomes (EO) is often highlighted as a case where military and security contractors have done well (Jones 2006, 363). EO, who were commissioned by the Sierra Leonean government in 1995 to fight against the Revolutionary United Front (RUF) rebels and create stability, fulfilled their contractual obligations and secured stabilisation in this war-torn country – at least in the short term – by forcing the RUF to the negotiation table (Francis 1999, 327). Military and security contractors' actions during the wars in Iraq and Afghanistan, however, were marked by gravely poor behaviour that often undermined the political and operational aims of the US military. This was most evident with the Nisour Square incident, where Blackwater military contractors shot at Iraqi civilians, killing 17, whilst escorting a US embassy convoy (Human Rights First Report 2008, 1). Politically, the incident illuminated a grave accountability gap in the attempt to

---

3   See https://www.theguardian.com/world/2019/jun/11/leaked-documents-reveal-russian-effort-to-exert-influence-in-africa (accessed 13 June 2019).

4   See https://thedefensepost.com/tag/pmc-wagner/ (accessed 17 September 2019).

5   See https://www.bbc.co.uk/news/world-africa-52571777 (accessed 13 June 2020).

hold the Blackwater employees accountable for their actions. Several avenues of legal prosecution, including local Iraqi law,[6] were impossible to follow (Chen 2009, 106), ultimately creating a 'political firestorm' in both Iraq and the US (Human Rights First 2010, 1).

Operationally, Iraqi militias did not differentiate between the US military and security contractors, leading to further attacks on the former as an outcome of the incident. The Iraqi Government also revoked Blackwater's license to operate in the country, making it harder for the State Department to perform its functions because of its heavy reliance upon the company for the provision of security details. Further, it is doubtful whether US and UK diplomats could operate in high-risk environments without the protective services provided by military and security contractors today (Cusumano and Kinsey 2019, chapters 1 and 3), testifying to an increasing reliance on these actors. Military and security contractors undertake many different activities across many different environments, usually concurrently with other remote warfare defence activities, such as using military drones to target high-value enemy commanders. Furthermore, it is important to note that in some cases the activities are linked, so that military and security contractors are often employed to maintain military drones as well as operate them.

The rationales behind utilising military contractors in remote warfare

Before discussing why states choose to use military and security contractors instead of relying on military personnel, it is worth engaging with some of the counterarguments to their use. Opponents of military and security contractors are quick to point out the role of the military is to protect the state, and that such an important responsibility should not be subject to commercial interests (Pattison 2010, 437–439). To put it differently, critics argue that it is dangerous to put a financial figure on national security because the price of failure is too high. Such critics also point out that contractors can leave their post at any time. Importantly, unlike military personnel, contractors cannot be ordered to stay and, if necessary, fight. Furthermore, if they are providing a critical service (for example, supporting satellite communication equipment), leaving could threaten the successful outcome of a military operation. Finally, while much has been said about the cost efficiency of using contractors, as opposed to military staff, the evidence is not conclusive (Stanger 2009, 94–98). Contractors also regularly take on roles today that in the past the military considered important because they allowed personnel to rest between operations. Without these postings, the military risk over-committing some of

---

[6]   Before Paul Bremer (leader of the Coalition Provincial Authority) left Iraq in December 2006, he issued 'Order 17,' which effectively ensured that the military contractors working in Iraq would be exempted from prosecution under Iraqi law.

their staff to operations that might lead to them leaving the services.

Today, it is often the case that military and security contractors are seen by states as nothing more than a practical answer to some of the challenging working conditions that are part of military operations. These challenging conditions may revolve around budgetary constraints, technology, risk aversion, or the operational weakness of local partners. Still, these conditions are not new, nor is the use of military contractors to resolve them. As Parrott (2012, 135) reminds us when talking about military enterprises in early modern Europe, 'No state had the resources or the organisation to create a significant army directly under its own funding or control, and the only issue was what level of private enterprise would be encouraged or allowed.' While the character of war has changed since then, many of the challenging conditions of warfare remain and the need to hire military contractors therefore persists. Continuing, Parrott (2012, 135, 177) explains that while the involvement of military enterprisers in early modern European warfare was tantamount to adopting a wasteful and ineffective military system, they were integrated into a much larger network of producers, suppliers, merchants and distributors, upon whose resources the entire logistical structure of the war effort depended. In effect, military enterprises offered states valuable resources, capabilities and capacity that they could not generate internally – a situation not unlike the one many modern militaries find themselves in today.

As noted, the rationales behind utilising military and security contractors appear to have changed little over the last 600 years. Today, the most common reason for employing contractors is to improve organisational efficiency by exposing the military to market forces (Uttley 2005). Since the 1980s, contractorisation[7] has been seen as a way of restructuring defence in an attempt to manage budgetary constraints more efficiently. It is also thought to be a more financially sustainable way of maintaining military equipment on operations by drawing defence contractors into the battlespace to support their equipment (Kinsey 2014, 5). Advancements in communications and weapon technology are also driving the contractorisation of large sections of the support element of military operations. For example, the rapid introduction of technically advanced weapon systems, such as unmanned aerial vehicles (commonly referred to as UAVs), has left the military ill-prepared to maintain and operate them and needing to use contractors to perform these tasks. Military and security contractors are also often better qualified and more experienced in maintaining sophisticated weapon platforms, especially when equipment is newly fielded (Camm 2012, 239). Finally, relying on contractors

---

[7]	Contractorisation refers to the outsourcing of publicly performed services to commercial actors through the use of legally binding contracts. These contracts determine the nature of the working relationship between the two organisations.

to generate strategic capabilities using technology avoids training military personnel in essential and costly skill sets (Heidenkamp 2012, 15). Thus, without military and security contractors, it would be impossible for countries such as the US and UK to wage high-technology warfare over the short term on the basis of the resources that could be mobilised or extracted by states and their direct agents, a situation not too dissimilar to that of seventeenth century Europe (Parrott 2012, 260).

A third rationale for using contractors in military operations is a reduction in political risks associated with soldiers returning home in body bags. Returning dead contractors rarely make the news, and, even if they do, their efforts are often not cherished by society in the same way as the efforts of state soldiers. While using military and security contractors is often done to minimise a loss of public support for unpopular conflicts by either satiating a risk-averse society or hiding the true costs of the military efforts, such an approach can pose several challenges. The use of military contractors means that states are one step removed from the action taken by these actors on the ground. Critics of the modern use of military and security contractors have noted that this is an infringement of public accountability due to the lack of transparency and knowledge about military contracting (Liu 2015, 84–89). This means that the future use of military contractors will have to reconcile the inherent need for discretion in this area with the demand for accountability for these actors and their actions, and general transparency about the use of military and security contractors in warfare.

The final reason behind why states turn to military and security contractors relates to their support of local partners. The conflicts in Syria and the Donbass region of Ukraine provide examples of how a state intervening in civil wars can utilise military contractors to enhance the military effectiveness of local partners without committing ground troops. Russia effectively delegated the risk of soldier deaths to the market for force in the hope that the population at home would remain silent over the legitimacy of the intervention. Such an approach is also financially attractive as the cost of hiring the contractors is often carried by the local partner. However, this practice also holds challenges for states. While outsourcing the teaching of defence doctrines, training and force design to contractors may reduce the political risk of deploying soldiers, it does not guarantee success on the battlefield. For example, Russian contractors fighting alongside local pro-Russian forces in the Donbass region have had little impact on the general outcome of the fighting (Isenberg 2018). Nevertheless, they helped Putin hide the true cost of Russian deaths from his people, something he could not have done if he had used soldiers, while pledging support to pro-Russian forces in the region. The same approach has been used in Syria, where Russian contractors are supporting local forces loyal to the government of Bashar al-

Assad (Gibbons-Neff 2018). However, unlike the Russian intervention in Ukraine, in Syria they have been more successful in training and supporting local forces fighting the Islamic State (Ibid.). Even so, such an approach may also send the wrong message to local forces, in that they are not valued highly enough to deploy regular soldiers to support them.

The future of military contracting in remote warfare

The future of military contracting and the use of private actors in remote warfare are intimately tied to specific practical, ethical, political, and legal considerations and concerns. Among these are concerns about the continued use of these actors alongside the adherence to core democratic values such as accountability, transparency and public consent. As discussed above, the legal prosecution in the advent of wrongdoings by individual military contractors is made difficult by their status as both *private* and *military*, and by the fact that they are usually grouped in *companies*. This can ultimately frustrate attempts to regulate and prosecute these actors (Liu 2015, 3). Furthermore, considerations about whether the military *should* and *could* maintain capabilities that are currently outsourced to private actors are central points of contention.

Today, it is unimaginable for states like the US or the UK to go to war without support from military and security contractors (Kinsey and Patterson 2012, 1), whether remotely or conventionally, and this is unlikely to change in the future. There are two primary reasons for this. Firstly, since the end of the Cold War militaries have struggled to recruit enough manpower into their ranks.[8] This shortfall is now being filled by contractors performing mundane but critical functions such as providing security, aviation and communication services. Secondly, the military no longer has enough specialist skill sets; for example, helicopter pilots (NAO 4 September 2019 Investigation into Military Flying Training, MoD, HC 2635), nor the range of skill sets needed for military interventions. For instance, they lack staff with the range of language skills to cover every country where an intervention may occur. The only way to get around this problem, therefore, is to hire contractors.

As noted above, military and security contractors are often employed on an *ad hoc* basis and not as part of a wider military strategy. This can pose practical challenges for future remote warfare efforts, as the success of military operations often rests upon a broad overall strategy where all

---

8   See https://www.armytimes.com/news/your-army/2019/03/14/after-2018s-recruiting-shortfall-it-will-take-a-lot-longer-to-build-the-army-to-500k/ (accessed 17 September 2019); https://assets.publishing.service.gov.uk/government/uploads/system/uploads/attachment_data/file/824753/201907-_SPS.pdf (accessed 17 September 2019).

components work together effectively. If use of military and security contractors happens on an informal and hidden basis, it can damage the potential for a successful operation. However, the use of military and security contracting also offers distinct opportunities for states. If military and security contracting is an accepted part of conducting military operations and is used in an active and conscious way, these actors can be used more openly and as a mindful part of future strategy. This will in turn affect how a country relates to its allies, and which future military operations we can participate in.

However, there still exists a widespread negative outlook on the use of military and security contractors in the theatre of war. This can, for example, be seen in media reports on these actors where the use of the word 'mercenary' is used to evoke a specific set of negative prejudices.[9] This perception is largely a hang-up from the 1960s and 1970s and memories of the 'dogs of war.' The phrase 'dogs of war,' originally uttered by Marc Antony in Shakespeare's *Julius Caesar* (Shakespeare, 2006, 3.1.273), has since been appropriated by scholars, commentators and journalists alike to describe the predominately white, battle-hardened mercenaries working on the African continent in various independence wars during the 1960s and 1970s (Frye 2005, 2622).[10]

The negative perception of mercenaries, which Anthony Mockler (1969, 13–24) argues has always been present,[11] still affects military and security contractors today. Joachim and Schneiker (2012, 365-267) argue that contemporary military and security contractors employ 'frame appropriation' techniques to escape the mercenary label, while Krahmann (2012, 345–346) argues that military and security contractors use different discursive strategies to construct themselves as viable and legitimate actors. As such, the close association of military and security contractors with a negative mercenary label still affects these actors today. This sentiment often ends the possibility of public debate about the modern use of military and security contractors – a debate that is wholly necessary if these actors are to be used effectively in the future.

While devising military strategy should not necessarily be a public activity, public debate about the inclusion of military and security contractors in

---

[9]   A recent article in *The Intercept* referred to Blackwater founder Erik Prince as a 'mercenary'- https://theintercept.com/2019/05/03/erik-prince-trump-uae-project-veritas/ (accessed 5 June 2019).

[10]   A popular example is the novel 'The Dogs of War' (1981 New York: Random House) by journalist Frederick Forsyth where he describes the conduct of an army of mercenaries working in a fictitious African country.

[11]   Sarah Percy argues that there is a norm against mercenary use (Percy 2007,1).

military operations must take place if liberal democratic states truly treasure ideals such as transparency and accountability. Transparency is one of the core tenets of liberal democracies in that it provides the citizenry with access to key information, thus enabling them to make informed decisions and participate fully in the democratic processes (Avant and Sigelman 2010, 236, 243). When access to information is restricted – as it often is in the area of military activity – it inhibits the proper participation in free and open debate with the consequence that policy decisions will be in the interest of the few, rather than the many. The lack of transparency thus prevents the citizenry from engaging fully and informedly in political discussions. Furthermore, the lack of access to reliable information about the level of military engagement can 'diminish the perceived human costs of war,' with severe ethical implications (Avant and Sigelman 2010, 256). Therefore, states must take the demand for greater transparency seriously and be open about which mechanisms are in place to ensure that military and security contractors fulfil their contract effectively and efficiently.[12]

Moving away from a reactionary outlook on military and security contractors also requires a hard look at our relationship with war and those who fight for us or at least support the ability to fight an adversary – it requires a completely new and more open approach to military and security contracting. Military and security contracting needs to be brought into the light where it is possible to acknowledge the vital and varied roles military and security

---

[12]   An example of a government department concealing their use of military contractors is the UK Foreign and Commonwealth Office (FCO), which employs contractors to train the Lebanese army. On the 5 March 2017, Dr Kinsey, one of the authors of this chapter, emailed a Freedom of Information Request (FOIR) to the department asking if it could confirm whether the FCO was funding a private security company to help train the Lebanese army. His request was turned down (REF: 0254-17). Dr Kinsey then asked for an internal review. The review upheld the exemption cited in the refusal notice. Dr Kinsey then contacted the Information Commission on the 6 June 2017 in order to complain about the FCO's handling of his request. The outcome of this request was to uphold the original exemption. Dr Kinsey appealed this decision with the First-tier Tribunal in November 2017. The Tribunal's decision, taken on the 13 February, was also to uphold the original exemption (REF: EA/2017/02830). Then, in February 2019, Dr Kinsey came across an FCO blog that mentioned UK military veterans have supported the training, mentoring and equipping of the Lebanese Armed Forces. As he believed these veterans could only be military contractors, Dr Kinsey put in another FOIR on 15 February 2019 and, on 10 April, this was confirmed by the FCO (REF: 0209-19). See https://blogs.fco.gov.uk/stories/uk-watch-towers-protecting-lebanon-from-daesh/ (accessed 3 June 2019). The company responsible for performing this role is Risk Advisory. The company has been working in Lebanon for the past decade. Even so, little is known about what they do and who is paying them. See https://www.riskadvisory.com/news/risk-advisory-on-sky-news-how-lebanons-borders-are-preventing-terrorists-reach-europe/ (accessed 13 June 2019).

contractors play in modern warfare, which, in turn, will enable a more frank discussion about how the future of military and security contracting should look. Part of this discussion involves understanding the actual roles military and security contractors undertake in conflict zones, as well as a greater understanding of the types of remote warfare a given state participates in. However, this debate needs to rest on a wider discussion of how war – both abstractly and practically – is viewed in civil society, and it requires an honest discussion of what it means to have a military today and how it can be used in the future. Nevertheless, debate can only happen if there is transparency about the degree and scale of military and security contracting. As mentioned above, one of the attractive aspects of military and security contracting is the ability to keep the practice out of the public eye – an aspect that is seemingly contradictory to democratic debate.

Conclusion

As discussed above, military and security contractors are now part of the conflict environment engaging in remote warfare, while undertaking lethal and non-lethal services for state militaries. The reasons for the growth in military and security contracted services are numerous. Nevertheless, the two most common are the military's inability to generate certain capabilities in-house, and that it is often financially cheaper to buy-in some capabilities than to use military personnel. Still, there are important ethical, political and legal concerns with military contracting. The most serious of these is the fear that military and security contracting will lead to a democratic deficit, where accountability, transparency and public consent are either ignored or quietly marginalised in favour of political and strategic expediency. Moreover, ignoring the wishes of the public is tantamount to the privatisation of foreign policy and the return to warfare by cabinet. Importantly, if military and security contractors are to provide states with remote warfare opportunities, then contractors must first be properly incorporated into the military's official force structure. However, this will entail integrating their activities into the broader strategic aims of government; a task that will not be easy, given the hostility towards them and their ability to allow states to hide their activities from public scrutiny.

References

Abrahamsen, Rita, and Williams, Michael C., 2011. *Security Beyond the State: Private Security in International Politics*. Cambridge: Cambridge University Press.

Avant, Deborah, and Lee Sigelman. 2010. 'Private Security and Democracy: Lessons from the US in Iraq', *Security Studies*, 19(2): 230–265.

Camm, Frank. 2012. 'How to Decide When a Contractor Source is Better to Use Than a Government Source.' In *Contractors & War: The Transformation of US Expeditionary Operations*, edited by Christopher Kinsey and Patterson, Malcolm H.

Chen, David H., 2009. 'Holding "Hired Guns" Accountable: The Legal Status of Private Security Contractors in Iraq.' *Boston College International and Comparative Law Review,* 32.

Cole, Matthew. 2019. 'The Complete Mercenary: How Eric Price used the rise of Trump to make an improbable comeback.' *The Intercept.* 3 May. Accessed 5 June 2019. https://theintercept.com/2019/05/03/erik-prince-trump-uae-project-veritas/

Cusumano, Eugenio, and Christopher Kinsey. 2019. *Diplomatic Security: A Comparative Analysis*. CA.: Stanford University Press.

Dunigan, Molly, et al. 2013. *Out of the Shadows: The health and Wellbeing of private Contractors Working in Conflict Environments*. US: The RAND Corporation.

Francis, David J., 1999. 'Mercenary Intervention in Sierra Leone: Providing National Security or International Exploitation?' *Third World Quarterly*, 20(2).

Frye, Ellen L., 2005. 'Private Military Firms in the New World Order: How Redefining 'Mercenary' can tame the "Dogs of War".' *Fordham Law Review,* 73.

Gibbons-Neff, Thomas. 2018. 'How a 4-Hour Battle Between Russian Mercenaries and US Commandos Unfolded in Syria.' *The New Your Times.* 24 May.

Hawks, Alison. 2014. *The Transition from the Military to Civilian Life: Becoming a Private Security Contractor after Military Service for US and UK Service Leavers*. PhD Thesis. King's College London.

Heidenkamp, Henrik. 2012. *Sustaining the UK's Defence Effort: Contractor Support to Operations Market Dynamics*. Whitehall Reports. London: RUSI. April.

Human Rights First. 2008. *Private Security Contractors at War: Ending the Culture of Impunity*. New York.

————. 2010. *State of Affairs: Three Years after Nisoor Square*. US, New York.

Isenberg, David. 2018. 'Putin's Pocket Army? The Rise of Russian Mercenaries in Syria.' *The American Conservative.* 15 February. Accessed 6 July 2019. https://www.theamericanconservative.com/articles/putins-pocket-army-the-rise-of-russian-mercenaries-in-syria/

Joachim, Jutta, and Schneiker, Andrea. 2012. 'New Humanitarians? Frame Appropriation through Private Military and Security Companies.' *Millennium: Journal of International Studies,* 40(2).

Jones, Clive. 2006. Private Military Companies as 'Epistemic Communities.' *Civil Wars,* 8(3–4).

Kinsey, Christopher. 2006. *Corporate Soldiers and International Security: The Rise of Private Military Companies*. New York: Routledge.

————. 2006b. 'Examining the Organisational Structure of UK Private Security Companies.' *Defence Studies*, 5(2): 188–212.

————. 2014. 'Transforming war supply: Considerations and rationales behind contractor support to UK overseas military operations in the twenty-first century.' *International Journal*. London.
DOI 10. 1177/0020702014542814, Canada: Sage.

Kinsey, Christopher, and Malcolm H. Patterson. 2012. *Contractors and War: The Transformation of US Expeditionary Operations*. Stanford, CA.: Stanford University Press.

Knowles, Emily, and Abigail Watson. 2018. *Remote Warfare: Lessons Learnt from Contemporary Theatres.* Remote Warfare Programme, Oxford Research Group.

Krahmann, Elke. 2012. 'From "Mercenaries" to "Private Security Contractors": The (Re)Construction of Armed Security Providers in International Legal Discourses.' *Millennium: Journal of International Studies,* 40(2).

Liu, Hin-Yan. 2015. *Law's Impunity: Responsibility and the Modern Private Military Company*. Oxford and Portland, OR: Hart Publishing.

Mockler, Anthony. 1969. *The Mercenaries*. London: Macdonald.

Moore, Adam. 2017. 'US Military Logistics Outsourcing and the Everywhere of War' *Territory, Politics, Governance,* 5(1): 5–27.

National Audit Office. 2019. Investigation into Military Flying Training. MoD, HC 2635.

Østensen, Åse Gilje. 2013. 'In the Business of Peace: The Political Influence of Private Military and Security Companies on UN Peacekeeping.' *International Peacekeeping,* 20(1): 33–47.

Pattison, James. 2010. 'Deeper Objections to the Privatisation of Military Force'. *The Journal of Political Philosophy,* 18(4): 425–447.

Parrott, David. 2012. *The Business of War: Military Enterprise and Military Revolution in Early Modern Europe*. Cambridge: Cambridge University Press.

Percy, Sarah. 2007. *Mercenaries: The History of a Norm in International Relations*. Oxford: Oxford University Press.

Phelps, Martha Lizabeth. 2016. 'Supporting the Troops: Military Contracting in the United States.' In *The Routledge Companion to Security Outsourcing,* edited by Joakim Berndtsson and christopher Kinsey. Oxford, Routledge: 9–19.

Shakespeare, William. 2006 [1599]. *Julius Caesar,* edited by Burton Raffel. New Haven, CT and London: Yale University Press.

Stanger, Alison. 2009. *One Nation Under Contract: The Outsourcing of American Power and the Future of Foreign Policy.* New Haven: Yale University Press.

Uttley, Matthew. 2005. 'Contractors on Deployed Military Operations: United Kingdom Policy and Doctrine.' US: Strategic Studies Institute.

5

Outsourcing Death, Sacrifice and Remembrance: The Socio-Political Effects of Remote Warfare

MALTE RIEMANN AND NORMA ROSSI

Late modern warfare is increasingly characterised by 'the technical ability and ethical imperative to threaten and, if necessary, actualise violence from a distance – with no or minimal casualties' (Der Derian 2009, xxi). The term remote warfare has been coined to capture this process where states and societies of the Global North are progressively distancing the effects of war. New technologies, such as drones, and actors, such as private military and security companies (PMSCs) and special forces, are a fundamental feature in enabling such types of warfare, and their importance has attracted increasing attention (Chamayou 2015). In this chapter, we focus on what Der Derian has referred to as the 'ethical imperative.' This imperative, we argue, underpins the commitment towards forms of remote warfare and actively shapes the direction and focus of the techniques it employs. In order to think about remote warfare, it is necessary to recognise the normative commitment that underpins this way of war. This is a commitment which emerges clearly from the definition of remote warfare as a series of methods and approaches, such as the use of proxies, special operations forces, PMSCs and drones, to 'counter threats *at a distance*' (Watts and Biegon 2017). The chapter focuses on the ethical imperative sustaining the process of distancing by looking at the normative commitment embedded within forms of remote warfare. We do so by exploring remote warfare's socio-political effects on intervening states, which so far has generated only limited attention from scholars.

Recent literature on remote warfare, or variously termed 'liquid warfare'

(Demmers and Gould 2018), 'surrogate warfare' (Krieg and Rickli 2018) and 'vicarious warfare' (Waldman 2018), has mainly focused on the very spaces and times in which remote forms of warfare are enacted. In this, the literature has moved its focus away from an analysis of remote warfare's legal and technical aspects (see Rae 2014; Boyle 2015), and towards the socio-political effects this form of warfare has on the everyday social realities of people living within the areas where remote warfare takes place. Studies have shown remote warfare's impact on the lived realities within theatres (Calhoun 2018), demonstrated how drone strikes undermine the legitimacy of states and governments at the receiving end of these interventions (Boyle 2013), exposed how PMSCs blur the distinction between civilians and combatants, extending the space of the battlefield and blurring its borders (Kinsey 2006), and highlighted how intervening states are increasingly privileging long-distance air strikes and the training of local forces over long term state-building processes with detrimental effects for local security (Kaldor 2012, 151–184; Knowles and Watson 2018). By exposing how remote warfare contributes to turning war into the permanent socio-political condition for people living within the vicinity of these interventions, this literature offers a powerful critique of this method of engagement. Indeed, remote warfare's socio-political effect of turning war into a permanent condition for underprivileged spaces and times makes remote warfare everything but remote. War rather becomes perpetually present in space and time as expressions such as 'everywhere war' (Gregory 2011) and 'forever war' (Filkins 2008) capture.

Remote warfare's socio-political effects on the states and societies from which it originates, however, have so far received only limited attention. This chapter turns to this overlooked aspect by analysing the socio-political effects the seeming absence of war has on the societies of the intervener. Our argument unfolds in three steps.

First, by focussing on the etymology of the term 'remote', we expose that remote not only entails a physical distancing but also encapsulates a specific normative commitment to temporalise the states in which remote interventions take place, framing them as morally backwards and thus paving the ground for military intervention. Second, we show that remote warfare challenges the traditional ways in which societies in the Global North have sustained their projects of nation-building through the production of a collective identity based on/in sacrifice (Kahn 2013; Taussig-Rubbo 2009). Third, we analyse the practice of military outsourcing as a tool of remote warfare. Specifically, we show how the outsourcing of death to private proxies exposes the ways in which neo-liberal states are renegotiating the very meaning of what it means to sacrifice for the collective identity that the nation has historically claimed to express.

The space and time of remote warfare

As argued above, remote warfare contains the ethical imperative to distance war. Indeed, the very act of distancing is hidden in plain sight within the very term itself: *remote* warfare. The etymology of 'remote' allows us to shed light on remote warfare's normative dimension by exposing that remote encapsulates both spatial as well as temporal distancing. Etymology is a useful tool in this regard as it uncovers the whole range of various meanings that a term can carry, thereby contributing 'to the understanding of the performativity of language in making the world in which "we" live in.' (Riemann 2014, 3).

Remote, deriving etymologically from the Latin adjective 'remotus' for 'distant in place, afar, set aside, removed' shows how the term expresses the spatial logics sustaining remote warfare (Castiglioni and Mariotti 1996, 1097). As such, remote in space signifies the commitment for distancing war from 'over-here', while simultaneously maintaining the possibility of fighting it 'over there.' Perpetuated by 9/11, elements of this spatial logic found expression in the Bush Doctrine's notion of pre-emption based on the proposition that, 'we will fight them over there, so we do not have to face them in the USA' (Bush 2007). This approach of fighting wars at a distance continued under the Obama administration's extension of the US drone programme and has further intensified since Trump took office (Rosenthal and Schulman 2018). Besides its spatial logic, the meaning of 'distance' entailed within the term remote warfare also contains a temporal quality. New technologies, for example, not only permit interveners to recede 'further in time and space from the target of military operations' (Ohlin 2017, 2) but also 'bring "there" here in near-real time' (Der Derian 2009, xxxi). However, the use of virtual technologies to conduct military operations from afar with near verisimilitude, is not the sole temporal aspect of remote warfare.

Remote in time, we argue, is also linked to imaginaries of underdevelopment, civilisational standards and ideas of backwardness that are often associated with the places in which remote warfare takes place. This temporal connotation is deeply embedded into the very term remote, even though contemporary English privileges the word's spatial dimension. Remote's etymology is again indicative, as the Latin *remotus* refers to '*distance in time*', but also '*different, adverse, alien*' (Zalli 1830, 492–493). Even in English, both the temporal as well as the aspect of difference, were included in its meaning until the nineteenth century. Samuel Johnson's (1828, 286) *A dictionary of the English Language* exemplifies this, as it defines remote as '1. Distant in time, not immediate, 2. Distant in Place … 4. Foreign … 6. Alien; not agreeing.' What we find within remote warfare, therefore, is what Barry Hindess (2007) has characterised as the 'temporalisation of difference', through which certain contemporaries and the spaces they inhabit are assigned to an anterior time. Moreover, subjects inside these 'backward' spaces are portrayed as morally bankrupt and fundamentally different in comparison to their contemporaries (Ibid., 325–326).

The term 'remote' thus hides in plain sight the ways in which subjects and spaces where remote interventions take place are constructed as backward and distant in time through the process of temporalisation. And it is precisely this temporalisation which makes these subjects and spaces 'targetable.' This is most visible in relation to discourses on fragile and failing states, which form the backdrop for most remote interventions (Fernández and Estevez 2017, 149; Watts and Biegon 2017; Waldmann 2017). Debates on these spaces deploy a variety of metaphors and characteristics to locate fragile states on a temporal scale in which these are variously defined as 'medieval' (Forrest 1994), belonging to a Hobbesian state of nature that precedes the social contract (Kaplan 1994) or simply 'pre-modern' (Cooper 2003). Such representations 'inferiorise difference by interpreting it as backwardness' and delegitimises these spaces 'through a comparison – explicit or implicit – with temporally more advanced identities' (Moreno 2015, 72). Furthermore, these 'discursive practices, based on a Eurocentric account, construct the "failed state" as deviant' thereby creating 'favourable conditions for interventionist practices' (Moreno 2015, 1).

Rita Abrahamsen (2005) observed the open-ended nature of these interventionist practices in the discursive change on fragile states that appeared after 9/11. Where previously 'development' and 'humanitarianism' were key terms of reference in debates on fragile states, these were gradually replaced with an insistence on categories of risk, fear and threat, that are in need of being continually contained to safeguard temporally advanced spaces (Ibid.). The spatial and temporal logics of remote warfare, therefore, contain the normative commitment of removing war from some privileged spaces and times even at the costs of turning war into a permanent social condition for underprivileged spaces and times. In doing so, remote warfare establishes a radical duality between spaces and times in which war is consistently present, and spaces and times from which it is removed. Put differently, from the perspective of societies in the Global North, the effects of 'being at war' are rendered invisible and its costs are largely placed on the societies that have become the object of remote forms of intervention.

Yet, the normative commitment of removing war from 'Western' societies is neither uncontested nor without consequences. First, because this attempt is consistently resisted. Terrorist attacks conducted in the 'West', for example, have often been framed as retaliatory actions to Western military interventions, including those under the label of remote warfare. For instance, on multiple occasions Islamic State justified attacks within Western societies as direct responses to what is happening in the theatres of remote warfare (Greenwald 2016). Second, remote warfare does not leave societies from which it originates untouched. Critical scholarship has been instrumental in exposing the profound political and legal effects that remote warfare has on liberal democracies, such as the lack of democratic accountability in the enactment of these wars (Baggiarini 2015; Chamayou 2015) and the increasing use of emergency/exceptional legislation (Neal 2010, 2015). Critical terrorism studies exposed the deep socio-political effects of remote warfare in Western states by raising awareness of the militarisation of domestic security and the use of techniques that travel from COIN 'abroad' to counterterrorism at 'home' (Owens 2015; Dunlap 2016; Sabir 2017). The ways in which Muslim communities in Western societies are increasingly the target of security practices, such as surveillance, stigmatisation and policing, is a case in point, (Awan 2011) suggesting that for some sections of the population in the 'West' remote warfare's effects are anything but remote. The reasons above highlight the importance of considering how the normative commitment of conducting remote warfare produces concrete socio-political effects within the societies from which war is supposedly removed. In the remaining part of this chapter, we turn our attention to how remote warfare affects a key component of the construction of modern statehood: the citizenship/sacrifice link (Hutchinson 2017).

Sacred soldier bodies and the citizenship/sacrifice link

Max Weber (2009, 78) famously defined the state as a 'community that successfully claims the monopoly of the legitimate use of violence within a given territory.' A sole focus on the physical/material aspects in Weber's definition, however, overlooks his engagement with the emotional foundations of political authority/community. In *Religious Rejections of the World and Their Directions*, Weber (2009, 335) highlights the important emotional foundations of the legitimation of force, arguing that 'the location of death within a series of meaningful and consecrated events ultimately lies at the base of all endeavours to support the autonomous dignity of the polity resting on force.' Sacrificial authority, therefore, underlies the state's political authority and power (Marvin 2014). Bargu (2014, 124) calls this emotional foundation of Weber's monopoly of violence, the *monopsony* of sacrifice. Monopsony derives from the ancient Greek *monos* (single) and *opsonia* (purchase). As such, '[b]uilding on Weber, we can say that the modern state is not only the sole provider of legitimate force; it is also the sole receiver of political self-sacrifice' (Bargu 2014, 124).

Nationalism links individual sacrifice to the state, implying 'a transfer of authority and meaning from God to originating peoples and their cultures' (Hutchinson 2017, 9). In war, this assumes an especially strong meaning, 'when cults of the national dead are potent, extolling that those who die will live forever in the memory of the nation' (Ibid.). Military remembrance rituals express this link by generating, in the words of Hutchinson (Ibid., 3–4), 'a sense of in-group commonality.' The year 2018, for example, saw nations around the globe commemorate the centenary of the end of the First World War (1914–1918).

At the centre of these commemorative events lay the remembrance of those that died, with a specific focus on military fatalities. London, for example, displayed parts of the gigantic artwork of 888,246 poppies that flooded the Tower of London in 2014, in which each poppy represented a fallen member of the British armed forces.[1] Such events form an integral part within the construction of national narratives, because it is the remembrance of those who passed that creates a sense of unity and national belonging (Marvin and Ingle 1996), which in turn forges a relational identity between citizen and state.

In the words of Jens Bartelson (1995, 189), 'the modern subject and the modern state are linked inside knowledge, and the concepts of nation and community are used to express their unity.' Nationalism, as David Campbell (1992, 11) has argued, therefore needs to be understood as one of the many ways through which the modern state pursues its legitimacy. Roxanne Doty (1996) argues similarly, asserting that state sovereignty is endorsed by, and finds expression in, national identity.

Military remembrance rituals thus have a constitutive function in the production and reproduction of sovereign claims and the creation of national identities. Specifically, commemorative rituals contribute to the state's ontological security. Ontological security differs from physical security by being 'not of the body but of the self, the subjective sense of who one is' (Mitzen 2006, 344). In this, ontological security is essential for the body politic as its 'capacity for agency' derives from it (Ibid.). Commemorative rituals are therefore crucial in establishing claims to political identity and authority, as they construct the constitutive link between self-sacrifice and a sense of collective identity. Thus, soldiers play a prominent role in the state's construction of political authority, as the idea of the nation, and with it the modern conception of citizenship, is intrinsically linked to the idea of soldiering as a prerequisite for citizen rights (Janowitz 1976; Millar 2015; Kier and Krebs 2010).

---

[1]   For pictures and a description of this installation visit: http://www.hrp.org.uk/tower-of-london/history-and-stories/tower-of-london-remembers/

The idea that military service represents a prerequisite for citizenship emerged from French Revolutionary thought (Janowitz 1976; Heuser 2010; Osman 2015). In this timeframe we must also situate the emergence of the soldier's death perceived as a sacrifice (Denton-Borhaug 2011; Riemann 2014; Baggiarini 2014; Baggiarini 2015). With regards to the French Revolution, Durkheim observed that a community's aptitude 'for setting itself up as god or for creating gods was never more apparent than during the first years of the French Revolution. At this time [...] under the influence of the general enthusiasm, things purely laical by nature were transformed by public opinion into sacred things: these were the Fatherland, Liberty, Reason' (Durkheim 2009, 116). This sacredness in due turn was then conferred to the actor, who swore an oath to protect these sacred abstractions. This actor was the citizen, who, only by becoming a soldier enrolled in the national armed forces, could defend the community that guaranteed his citizenship. It is, however, not the act of defending this abstraction, but rather dying in its defence, which provides the nation and consequentially state sovereignty with a veneer of legitimacy and political authority. Paul Kahn (2010, 205) expresses this vividly: 'We maintain the nation by sacrificing the sons.' National identity, citizenship and sacrifice are thus intrinsically linked (Baggiarini 2015), and, as such, sacrifice plays a key function in the constitution of political authority. The historical link between citizenship and sacrifice, however, is increasingly challenged by outsourcing practices (Riemann, 2014; Baggiarini, 2015).

Military outsourcing and the absence of death

One of the central elements of remote warfare involves shifting the burden of risk and responsibility onto others thereby increasingly externalising the burdens of war (Krieg 2016). This is by no means to say that such practices are without historical precedents. Barkawi (2010) cautions us to be aware of the international context of state-force-territory relations that sustain the nation-state centric monopoly on violence. Subaltern agents like colonial soldiers, for example, were not only used to fight in European wars, but also used to police the vast European colonial empires (Barkawi 2010). However, remote warfare intensifies these long-term tendencies, as 'Western' societies are increasingly shifting the burdens of war onto external actors, while simultaneously removing the experience of battle from their own nationals. While colonial forces were used to augment 'Western' forces in both World Wars, 'Western' forces were nonetheless engaged in fighting and dying. Today's wars, such as those that fall under the remote warfare label, show, however, a decreasing commitment of 'Western' societies to accept casualties and consequentially war. To capture this change, several scholars in the last two decades have argued that Western societies have entered a 'post-heroic age' (Lutwack 1995; Coker 2002). Although this notion drew extensive criticism (Frisk 2017), societies of the Global North are increasingly contracting out security tasks to an assortment of proxy actors beyond the regular armed forces to shield their own societies from the effects of war, while continuing to engage militarily abroad (Bruneau 2013; Mumford 2013). PMSCs fulfil a key function in this regard by enabling states to fight war remotely in a fashion that obfuscates the very presence of war (Schooner and Swan 2012). Media reporting on contractor fatalities exemplifies this point.

While every regular military fatality is extensively covered in the press, contractor deaths receive limited attention. The *Washington Post's* website 'Faces of the Fallen' is a case in point (Washington Post n.d.). This website, 'not only identifies deceased soldiers, but humanizes each loss with a photograph, biographical information, and a description of each service member's final action' (Schooner and Swan 2010, 16). But information about contractor fatalities appears to be of no particular interest to societies that hire their services, as the 'faces' of fallen contractors are omitted from this website. A news story that hit the American media in late summer 2004 confirms this. It stated that US casualties had passed the 1000 killed in action mark, putting a great deal of pressure on the Bush administration. What this story missed, however, was the blunt reality that such figures had long been passed, if contractor deaths would have been included (Singer 2004, 10). With regards to the theatres in Iraq and Afghanistan, Schooner and Swan observed in 2010 that 'contractor deaths now represent over 25 percent of all US fatalities' in those conflicts (Schooner and Swan 2010, 16). But it is more than possible that contractor fatalities are far higher, since there is no indication that non-US deaths have been tracked with any reliability. Schooner and Swan (2012, 3) as such conclude that, '[o]n the modern battlefield, contractor personnel are dying at rates similar to – and at times in excess of – soldiers.'

Nevertheless, contractor casualties go unnoticed. As Avant and Sigelman (2010, 256) note: 'There is no running count of contractor deaths on the network news or on the DOD website. Photos of PMSC personnel who have died in Iraq are not part of the "honour roll" flashed across the screen at the end of the *PBS News Hour*.' Some of the effects the non-recognition of contractor deaths produces, have already been pointed out. As contractor casualties often escape public attention, they shield policymakers from negative press (Avant and Sigelman, 2010, 243–249; Schooner 2008, 78–91), while simultaneously lowering 'the political and financial costs of intervention by desensitizing home populations' (Porch 2014, 700) to reduce possible public opposition and circumvent public oversight (Knowles and Watson 2017).

The externalisation of the burdens of war to private contractors, we argue, not only provides potential savings to the state in hiding the true costs of war, but also poses a very real challenge to the state's political authority, as the next section shows.

Private military corps and the relocation of sacrifice

Having elaborated on the importance of sacrifice in the construction of sovereign claims and how states are increasingly outsourcing sacrifice, we now turn to the effects that the increasing reliance on seemingly non-sacrificial actors in pursuit of remote forms of warfare has on the political authority of states. Three effects stand out in this regard.

First, by removing death from the equation of war, remote warfare weakens the relationship between citizenship, sacrifice and national identity. As the above analysis has shown, through the commemoration of particular soldier bodies, the state is able to express the unity of particular citizens living within the shared territorial confines of the state. The soldier's dead body is therefore a powerful tool that expresses the unity of man and state articulated in terms of national identity grounded in sacrifice. Reliance on non-sacrificial actors threatens to sever this unity as their profane deaths do not generate the necessary collective practices of commemoration, which 'secure the unity of the "imagined (national) community", and its associated narratives and rituals, in the face of sometimes acute social divisions' (Ashplant 2000, 263).

Second, by rendering death invisible through the increasing practice of outsourcing sacrifice, not only is the very national identity of citizens threatened but also the very institution of the state itself, as sacrifice lies at the heart of the polity resting on force. In the words of Carolyn Marvin and David Ingle (1996, 4); 'Without the memory of blood sacrifice, the nation state cannot exist, or at least, not for long.' Or, put differently by Paul Kahn (2011, 153), 'without sacrifice, no sovereign.' The potential savings for states conducting remote warfare via outsourcing practices, expressed in blood and treasure, therefore, bear significant overlooked costs in relation to the construction of political authority.

Third, and most significantly, though remote warfare increasingly omits deaths from public attention, sacrifice and consequentially sovereignty, are not disappearing but rather relocated. At first sight, contractors could be framed as conforming to Agamben's articulation of *homo sacer*, as actors who can be killed but not sacrificed (Nikolopoulou, Agamben and Heller-Roazen 2007). However, although contractor deaths lack the state sanctioned component of sacrifice, it would be misleading to conceptualise these actors as *homo sacer*.

Instead of an absence of sacralisation, we rather find a relocation and rearticulating of sacrifice. Taussig-Rubbo (2012) identified initial points of this rearticulation in his analysis of the military medal system in which medals, like the US Purple Heart, function as a public honour that recognises sacrifice. Initially, the award of these had been exclusively restricted to members of the armed forces, but a privatised economy of commemoration is beginning to emerge. In 2008, for example, Blackwater introduced the Worldwide Defense of Liberty Medal which recognised the sacrifices of killed or wounded contractors, and the US government made contractors eligible for public honour as civilians (Taussig-Rubbo 2009). However, the 'deaths may be called "sacrifices" and recognised as deaths in the name of the nation, but the ceremonies where those awards are given are often private events and exclude the media' (Taussig-Rubbo 2012, 316). As such, both state and private sector recognition 'share an awkwardness in being neither public nor private events' (Taussig-Rubbo 2009, 124). The 'awkwardness' of this newly emerging privatised and state sanctioned medal system, we argue, has the function of re-designing the state and inscribing the logic of the market within it. Blackwater's ability, for example, to insist in a court case in 2007 that it was both, a private corporation as well as part of the sovereign body is a case in point (Taussig-Rubbo 2009, 134–135). Remote warfare thereby moves the site of sovereignty rather than undermining it. It is this commemorative aspect which distinguishes PMCs from other non-human means aimed at making war remote, such as, for instance, drones.

Baggiarini (2015, 130) has noted that the use of drones constitutes a 'logical extension' to the rationality of military privatisation in the quest for a 'bloodless' war on the side of the 'West.' She makes the valid point that the privatisation of war and drones respond to the same quest of removing the effects of war from the societies and the political bodies from which they originate, severing sacrifice from the body politic. Yet the socio-political effects of military privatisation and the use of drones are rather different; simply put, while drones cannot die, private contractors can. Instead of an eradication of death we rather find a relocation of death. Underplaying this fundamental difference risks ignoring the distinct socio-political effects that the displacement of death has on state sovereignty. While in the case of drone strikes the sacrificial component is removed, privatisation relocates it from the state to the market.

Conclusion

This chapter began by engaging with the normative commitment hidden in plain sight within the term 'remote warfare.' Remote warfare's normative commitment is the attempt to remove war from certain privileged spaces and times, even at the price of maintaining a perpetual and limitless condition of war elsewhere. While definitions such as 'everywhere war' and 'forever war' are effective in exposing how remote warfare contributes to extending war in time and space, these terminologies risk overlooking the centrality of the normative commitment to remove war from privileged spaces and times. We demonstrated this normative commitment through an analysis of the very etymology of the word remote, which implies the commitment to both spatial and temporal distancing. Remote Warfare, thus, works as an expression of a radical duality, in which war must be removed from the space and time of the self, while relocating it into the space and time of the 'Other.'

After exposing this normative commitment, we also argued that the attempt of removing war has important socio-political effects on the states and societies which wage war remotely. We explored these effects by analysing how the increasing use of private military and security contractors is an attempt to outsource death and render it invisible. We argue that this process undermines the link between the state and its citizens expressed through the imaginary form of the nation-state, in which the exceptional and com-memorated sacrifice of the soldier fulfils a central constitutive role for claims to state sovereignty. But this differs from the use of unmanned drones because the outsourcing of death to private contractors does not eliminate sacrifice, it only displaces it. This on-going process of the displacement of death from state to market deserves further investigation.

References

Abrahamsen, Rita. 2005. 'Blair's Africa: The Politics of Securitization and Fear.' *Alternatives: Global, Local, Political*, 30(1): 55–80.

Agamben, Giorgio. 1998. *Homo Sacer. Sovereign Power and Bare Life*. Stanford: Stanford University Press.

Ashplant, Timothy G., 2000. 'War commemoration in Western Europe: Changing Meanings, divisive loyalties, unheard voices.' In *The Politics of War Memory and Commemoration,* edited by Timothy G. Ashplant, Graham Dawson and Michael Roper. London and New York: Routledge: 263–272.

Avant, Deborah, and Lee Sigelman. 2010. 'Private Security and Democracy: Lessons from the US in Iraq.' *Security Studies*, 19(2): 230–265.

Awan, Imran. 2011. '"I am a Muslim not an Extremist.": How the Prevent Strategy has constructed a 'Suspect' Community.' *Politics and Policy*, 40(6): 1158–1185.

Baggiarini, Bianca. 2014. 'A Re-examination of Private Military and Security Companies.' *St Antony's International Review*, 9(2): 9–23.

———. 2015. 'Drone Warfare and the Limits of Sacrifice.' *Journal of International Political Theory*, 11(1): 128–144.

Bargu, Banu. 2014. *Starve and Immolate. Starve and Immolate*. Columbia University Press: New York.

Barkawi, Tarak 2010. 'State and Armed Force in International Context.' In *Mercenaries, Pirates, Bandits, and Empires: Private Violence in Historical Context*, edited by Alejandro Colás and Bryan Mabee. New York: Columbia University Press: 33-54.

Bartelson, Jens. 1995. *A Genealogy of Sovereignty*. Cambridge: Cambridge University Press.

Boyle, Michael J. 2013. 'The Costs and Consequences of Drone Warfare.' *International Affairs*, 89(1): 1–29.

———. 2015. 'The legal and ethical implications of drone warfare.' *The International Journal of Human Rights*, 19(2), 105-126.

Bruneau, C. Thomas. 2012. 'Contracting Out Security.' *Journal of Strategic Studies*, 36(5): 638–665.

Bush, George W. 2007. *President Bush Addresses the 89th Annual National Convention of the American Legion*. 28th August. https://georgewbush-whitehouse.archives.gov/news/releases/2007/08/20070828-2.html

Calhoun, Laurie. 2018.'Totalitarian Tendencies in drone strikes by states.' *Critical Studies on   Terrorism*, 11(2): 357–375.

Castiglioni, Luigi, and Mariotti, Scevola. 1996. *IL*, 3rd edition. Rome and Milan: Loescher.

Campbell, David. 1992. *Writing Security*. Minneapolis: University of Minnesota Press.

Chamayou, Grégoire. 2015. *A Theory of the Drone*. New York: The New Press.

Coker, Christopher. 2002. *Waging War without Warriors? The Changing Culture of Military Conflict*. Boulder: Lynne Rienner.

Cooper, Robert. 2003. *The Breaking of Nations: Order and Chaos in the Twenty First Century.* London: Atlantic Books.

Demmers, Jolle, and Lauren Gould. 2018. 'An Assemblage Approach to Liquid Warfare: AFRICOM and the "Hunt" for Joseph Kony.' *Security Dialogue*, 49(5): 364–381.

Denton-Borhaug, Kelly. 2011. *US War Culture, Sacrifice and Salvation*. London and Bew York: Routledge.

Der Derian, James. 2009. *Virtuous War. Mapping the Military-Industrial-Media-Entertainment-Network*. New York: Routledge.

Doty, Roxanne. 1996. 'Sovereignty and the Nation: Constructing the Boundaries of National Identity.' In *State Sovereignty as Social Construct*, edited by Thomas Bierstecker and Cynthia Weber. New York: Cambridge University Press: 121–147.

Durkheim, Emile. 2009. 'The Social as Sacred.'In *Introducing Religion: Readings from the Classic Theorists*, edited by Daniel Pals. Oxford: Oxford University Press.

Fernández, Marta, and Paulo Estevez. 2017. 'Silencing Colonialism: Foucault and the International.' In *Foucault and the Modern International: Silences and Legacies for the Study of World Politics*, edited by Philippe Bonditti, Didier Bigo and Frédéric Gros. New York: Palgrave MacMillan: 137–54.

Filkins, Dexter. 2008. *The Forever War*. New York: Alfred A. Knopf.

Forrest, Joshua.1994. 'An Asynchronic Comparison: Weak States in Post-Colonial Africa and Medieval Europe.' In *Comparing Nations: Concepts, strategies, substance*, edited by Mattei Dogan and Ali Kazancigil. Oxford: Blackwell. 260–296.

Frisk, Kristian. 2017. 'Post-Heroic Warfare Revisited: Meaning and Legitimation of Military Losses.' *Sociology*, 52 (5): 898–914.

Greenwald, Glenn. 2016. 'The deceptive debate over what causes terrorism against the West.'*The Intercept*. 16 January. https://theintercept.com/2016/01/06/the-deceptive-debate-over-what-causes-terrorism-against-the-west/

Gregory, Derek. 2011. 'The everywhere war.' *The Geographical Journal,* 177(3): 238–250.

Heuser, Beatrice. 2010. *The Evolution of Strategy: Thinking War from Antiquity to the Present*. New York: Cambridge University Press.

Hindess, Barry. 2007. 'The Past is another Country.' *International Political Sociology,* 1(4): 325–338.

Hutchinson, John. 2017. *Nationalism and War.* Oxford: Oxford University Press.

Janowitz, Moris. 1976. 'Military Institutions and Citizenship in Western Societies.' *Armed Forces and Society* 2(2): 185–204.

Johnson, Samuel. 1828. *A Dictionary of the English Language*. Heidelberg: Joseph Engelmann.

Kahn, Paul W. 2011. 'Criminal and Enemy in the Political Imagination.' *The Yale Review*, 99 (1), 148–167.

———. 2010. *Out of Eden: Adam and Eve and the Problem of Evil* Princeton: Princeton University Press.

———. 2013. 'Immagining warfare.' *European Journal of International Law*, 24(1): 199–226.

Kaldor, Mary. 2012. *New and Old Wars: Organized Violence in a Global Era* 3rd Edition. Cambridge: Polity Press.

Kaplan, Robert. 1994. 'The Coming Anarchy.' *The Atlantic Monthly*, 273(2): 44–76.

Kier, Elizabeth and Ronald R. Krebs. 2010. 'Introduction: War and Democracy in Comparative Perspective.' In *War's Wake: International Conflict and the Fate of Liberal Democracy*, edited by Elizabeth Kier and Ronald Krebs. Cambridge: Cambridge University Press.

Kinsey, Christopher. 2006. *Corporate Soldiers and International Security*. London: Routledge.

Krieg, Andreas. 2016. 'Externalizing the burden of war. The Obama Doctrine and US Foreign Policy in the Middle East.' *International Affairs*, 92(1): 97–113.

Krieg, Andreas, and Jean Marc Rickli. 2018. 'Surrogate Warfare: The Art of War in the 21st Century?' *Defence Studies*, 18(2): 113–130.

Knowles, Emily and Abigail Watson. 2017. *All Quiet on the ISIS Front: British secret warfare in an information age*. London: Oxford Research Group.

————. 2018. 'Remote Warfare. *Lessons learnt from contemporary theatres*. London: Oxford Research Group.

Luttwak, Edward N. 1995. 'Toward Post-Heroic Warfare.' *Foreign Affairs*, 74(3): 109–22.

Marvin, Carolyn and David Ingle W. 1996. 'Blood Sacrifice and the Nation: Revisiting Civil Religion.' *Journal of American Academy of Religion,* 64(4): 767–780.

————. 2014. 'Religion and Realpolitik: Reflections on Sacrifice.' *Political Theology*, 15(6): 522–535.

Millar, Katherine M., 2015. 'Death does not become her: An examination of the public construction of female American soldiers as liminal figures.' *Review of International Studies*, 41(4): 757–779.

Mitzen, Jennifer. 2006. *'Ontological Security in World Politics: State Identity and the Security Dilemma.' European Journal of International Relations, 12(3): 341–370.*

Moreno, Marta Fernandez. 2015. 'Reversing Polarities: Anarchical (Failed) States versus International Progress.' *Brazilian Political Science Review*, 9(3): 64–87.

Mumford, Andrew. 2013. 'Proxy Warfare and the Future of Conflict.' *The RUSI Journal*, 158(2): 40–46.

Nikolopoulou, Kalliopi, Giorgio Agamben, and Daniel Heller-Roazen. 2007. 'Homo Sacer: Sovereign Power and Bare Life.' *SubStance*, 93(29): 124–131.

Ohlin, Jens David. 2017. 'Introduction'. In *Research Hadbook on Remote Warfare*, edited by Jens David Ohlin. Cheltenham: Edward Elgar Publishing.

Osman, Julia. 2015. *Citizen Soldiers and the Key to the Bastille*. New York: Palgrave Macmillan.

Porch, Douglas. 2014. 'Expendable soldiers' *Small Wars and Insurgencies*, 25: 696–716.

Rae, James DeShaw. 2014. *Analyzing the Drone Debates: Targeted Killing, Remote Warfare, and Military Technology*. New York: Palgrave MacMillan.

Ricks, Thomas E. 2011. 'Contractor deaths surpass US Military losses in both Iraq and Afghanistan.' *Foreign Policy*. http://ricks.foreignpolicy.com/posts/2011/02/09/contractor_deaths_surpass_us_military_losses_in_both_iraq_and_afghanistan

Riemann, Malte. 2014. 'Conceptualising the Dichotomy between Private Military Contractors and Soldiers amis 'Society'.' *Political Perspectives,* 8(3): 1–15.

Rosenthal, Daniel J., and Loren Dejonge Schulman. 2018. 'Trump's Secret War on Terror', *The Atlantic*. 10 August. https://www.theatlantic.com/international/archive/2018/08/trump-war-terror-drones/567218/

Scheipers, Sybille, ed. 2014. *Heroism and the Changing Character of War Toward Post-Heroic Warfare?*. Basingstoke: Palgrave MacMillan.

Schooner, Steven L., and Collin D. Swan. 2012. 'Dead Contractors: The Un-Examined Effect of Surrogates on the Public's Casualty Sensitivity.' *Journal of National Security Law and Policy*. https://ssrn.com/abstract=1826242

———. 2010. 'Contractors and the Ultimate Sacrifice.' *Service Contractor*, 2003 (September): 16–18.

Singer, Peter W., 2004. 'The Private Military Industry and Iraq: What have we learned and where to next?.' Geneva: DCAF Policy Papers.

Taussig-Rubbo, Mateo. 2009. 'Outsourcing Sacrifice: The Labor of Private Military Contractors.' *Yale Journal of Law and Humanities*, 21(1): 103–166.

————. 2012. 'The Value of Valor: Money, Medals and Military Labor.' *North Dakota Law Review*, 88: 283–320.

Waldman, Thomas. 2018. 'Vicarious Warfare: The Counterproductive Consequences of Modern American Military Practice.' *Contemporary Security Policy*, 39(2): 181–205.

Washington Post: 'Faces of the Fallen.' Accessed 17 September 2018. http://apps.washingtonpost.com/national/fallen/

Watts, Tom, and Rublick Biegon. 2017. 'Defining Remote Warfare: Security Cooperation.' London: Oxford Research Group.

Weber, Max. 2009. 'Politics as a vocation.' In *From Max Weber: Essays in sociology*, edited and translated by Hans H. Gerth and Charles W. Mills. New York: Oxford University Press: 77–128.

————. 2009. 'Religious Rejections of the World and Their Directions.' In *From Max Weber: Essays in Sociology,* edited and translated by Hans H. Gerth and Charles W. Mills. New York: Routledge: 323–360.

Zalli, Casimiro. 1830. *Dizionario Peimontese, Italiano, Latino e Francese*. Carmagnola: Tipografia di Pietro Barbié.

6

Remote Warfare in the Sahel and a Role for the EU

DELINA GOXHO

'The glass is half full, it's complex and we have a lot to do, but I'm convinced we are on the right track' remarked French Defence Minister Florence Parly at the Munich Security Conference on 16 February 2019. [1] She added that she believed the French military presence in the G5 Sahel countries (Mauritania, Mali, Chad, Niger and Burkina Faso) will improve the security situation in the Sahel, a region which has, for some time, been a prominent theatre of intervention. In December 2012, French troops intervened in Mali to stop Islamist militants advancing on the capital Bamako, firstly through Opération Serval and then later with Opération Barkhane (as of 2014). Islamist groups had gained control over the northern part of Mali, capitalising on the instability caused by the Libyan civil war in the region. Opération Serval succeeded in its efforts to recapture territory. Opération Barkhane was then launched to provide long-term support to the wider region and prevent 'jihadist groups' from regaining control (Bacchi 2014). In the past few years, however, Burkina Faso, Mali and Niger have been suffering some of the deadliest attacks on record, as the area is being ravaged by tribal conflict and terror attacks (Chambas 2020).

On 3 April 2019, the Islamic State's Amaq Agency released its first video footage of an alleged attack against French forces in Mali on the border with Niger (Weiss 2019). At the 2019 Munich Security Conference, Foreign Minister of Burkina Faso, Mamadou Alpha Barry, also lamented increasing instability in the region, stating that the money promised to the G5 Sahel

---

[1]   Opening speech of the Munich Security Conference. Retrieved from https://www.defense.gouv.fr/english/salle-de-presse/dossiers-de-presse/discours-de-florence-parly-en-ouverture-de-la-munich-security-conference

force is yet to be disbursed.[2] France has kept about 4,500 troops and pushed for the creation of a force made up of soldiers from the G5 group to combat jihadist extremism. In addition to the lack of resources, the G5 Force's impact has also been reduced due to poor coordination amongst the five African countries (French Ministry of Defence 2019).

Remote warfare[3] conducted by Western forces is shifting its focus to the Sahel and as European states try to rely less on the US security apparatus, old legal challenges – especially those concerning armed drones and remote warfare more broadly – are emerging in new territories. This places a particular strain on local communities in the Sahel, who are kept in the dark about operations in their country. This chapter discusses the use of remote warfare in the Sahel and the problems it creates. The chapter then explores the potential avenues for peace in the region. In particular, the chapter argues why the European Union (EU) is best placed to be a peace broker in the Sahel.

Remote warfare in the Sahel

On 17 November 2018 at around 1:00 am (Brussels time), French Defence Staff reported that Niamey air base in Niger lost contact with a Reaper drone belonging to the Barkhane force, which was returning to base. The drone crashed in a desert area and no casualties were reported (DefPost 2018). After the news broke out, the French and European public acknowledged the existence of French remote warfare in the region (see VOA Africa 2018; DefenceWeb 2018; Le Figaro 2018). As of July 2018, four French Reaper drones have joined the airbase in Niamey, in order to increase Opération Barkhane's capabilities and in 2020 six more will be joining the mission (Cole 2018). In addition, France has now armed and is using its drones, while awaiting the development of the European project Eurodrone, which would also equip Italian and German forces and should be operational by 2025 (Charpentreau 2018).

---

[2]   The entire discussion can be found here https://www.securityconference.de/en/media-search/s_video/parallel-panel-discussion-security-in-the-sahel-traffick-jam/s_filter/video/s_term/Panel%20Discussion%20The%20Syrian%20Conflict%3A%20Strategy%20or%20Tragedy%3F%20Conference%20Hall/

[3]   By remote warfare, I mean the definition given by Emily Knowles and Abigail Watson in *Remote Warfare: Lessons Learned in Contemporary Theatres* (Oxford Research Group, June 2018): 'a form of intervention which takes place behind the scenes or at a distance rather than on a traditional battlefield, often through drone strikes and air strikes from above, with Special Forces, intelligence agencies, private contractors, and military training teams on the ground.'

In September 2017, Italy and Niger also signed an agreement to develop bilateral cooperation on security matters. It was believed that the agreement would only deal with migrant influxes, but it appears that the Italian defence company Leonardo will also benefit from the agreement, as revealed by a Freedom of Information Act in February 2019 (Labarrière 2019). This type of agreement does not need to be ratified and is not subject to parliamentary scrutiny, making it easier for the Italian Government to conduct security operations in the Sahel, without having to ask for parliamentary approval.

The Italian mission will be based in Niamey, within the US airbase and had initially been blocked by France, in a dispute with Rome over influence in the region (Negri 2018). Another aligned mission is the UN peacekeeping mission MINUSMA (United Nations Multidimensional Integrated Stabilisation Mission in Mali), made up of about 10,000 troops and 2,000 police officers. Finally, Germany is also stepping up its engagement in the security sector. In 2018 Burkina Faso became a partner country of the German training initiative, in order to help build capacities within the police and the gendarmerie. 'We will develop this further to embrace equipment and will provide about ten million euros to this end. We will also offer advisory services to be provided by the Bundeswehr,[4] also of the order of seven to €10 million,' pledged Angela Merkel in her visit to Ouagadougou, the capital of Burkina Faso In 2019.[5] In Niamey, she stated that assistance is being provided, especially with regard to training Niger's armed forces with 'about €30 million invested recently.'[6]

The EU, as a whole, is increasing its presence in the region as well. Aside from supporting the G5 Sahel countries in political partnership and through development cooperation, the EU is also providing support for security and stability through the provision of €147 million to establish the African-led G5 Sahel Joint Force through its three Common Security and Defence Policy missions: EUCAP Sahel Niger, EUCAP Sahel Mali and EU Training Mission (EUTM) in Mali. The latter, in particular, falls within the definition of security force assistance and partner capacity building as it provides military training to Malian Armed Forces. EUTM was deployed in March 2013 with the aim of restructuring the Malian military and improving the general security sector reform in the country.

In addition, since summer 2017, the EU launched a regionalisation process of Common Security and Defence Policy (CSDP) action in the Sahel, in order to both combine military and civilian spheres and bring Sahel countries closer to

---

[4]   This refers to German Armed Forces.

[5]   Retrieved from German Federal Government website https://www.bundesregierung. de/breg-en/news/stepping-up-cooperation-with-the-sahel-region-1605870

[6]   Ibid.

each other on security matters.[7] Such work is part of a wider effort on the part of European states to conduct operations in the region remotely. As for the US presence, Niger Air Base 201 in Agadez (Damon, 2017), a future hub for armed drones and other aircraft, is now operational. Air Base 201, a compound of three large hangars in the middle of the desert, is twice the size of Agadez itself (Maclean and Saley 2018) and houses the US armed drone mission in Niger that currently operates out of Niamey.

The US presence in the Sahel has increased considerably in the past few years. The Tongo Tongo ambush in Niger in October 2017, where 4 US and 5 Nigerien soldiers were killed by Islamic State in the Greater Sahara (ISGS) fighters, has changed the appearance of US engagement in the region, unveiling the nature of US shadow war in the Sahel, much like the crash of the Barkhane Reaper drone did for Europe. From 2002, the US has been conducting training missions for local forces to equip them to fight against Boko Haram, al-Qaeda in the Islamic Maghreb (AQIM) and other groups such as Jam'at Nasr al-Islam wal-Muslimin (JNIM), ISGS, Ansar al-Din, Ansar al-Islam.[8] Even though the rhetoric says that they wish to maintain a light footprint in Africa, US forces are certainly increasing their presence in the Sahel, albeit in a different way. Right after the Tongo Tongo ambush, Niger authorised US armed drone presence on its territory and the US began the construction of its drone camp in Agadez, a more central location in the region which would allow for better control over a larger swathe of territory.

The dangers of remote warfare in the Sahel

The fears associated with remote warfare could be largely grouped in internal and external. Internally, there is a real or perceived lack of ownership and a rise in conspiracy theories. Externally, there is an evident lack of public scrutiny over light footprint warfare, as will be mentioned later, and the danger of blowback. These fears derive directly from the hidden nature of remote warfare. The Nigerien Government has welcomed the presence of US troops, as long as they contribute to the eradication of terrorist activity in the country, but civil society in Niger appears distrustful of such a presence. A report by *The Guardian* in 2018 states that foreign military presence has had negative impacts on freedom of speech and many opposition leaders have lamented the lack of parliamentary oversight whenever foreign presence is authorised (Maclean and Saley 2018). The fear is that Niger will increasingly become a hub for geopolitical interests of great powers, which could lead to tougher treatment of dissent internally (Ibid.).

---

[7]   See European External Action Service Factsheet on G5 Sahel : https://eeas.europa.eu/sites/eeas/files/factsheet_eu_g5_sahel_0.pdf

[8]   For a map of armed groups in the region, see Lebovich (2019) https://www.ecfr.eu/mena/sahel_mapping

In addition, Niger spends about 21 percent of its budget on defence, which for a poor country represents a large percentage of its revenues (Bailie 2018). The securitisation of the Nigerien political sphere is viewed as a way to harness support for a government that would otherwise receive less approval. As for the legality of foreign powers presence, this does not depend on Nigerien parliamentary approval. Neither the US nor Niger are revealing the details of their cooperation. Niger's authorities state that these are not 'defence agreements', as Niger is purely a logistical hotspot. It therefore comes as no surprise that the Nigerien public are concerned. A CIA official interviewed during a visit to Niamey in July 2019 reported that whenever a strike is launched from the US' Niger Air Base 201 near Agadez, 'a CIA commander sends a WhatsApp text to his Nigerien counterpart, it is a gentlemen's agreement.' It would be hard to call this parliamentary oversight. It appears that the defence of Nigerien territory is ongoing. After the Tongo Tongo ambush in June 2018, French and US Special Forces took part in a fight against militants next to the Libyan border.

The evolution of the conflicts in the region points towards a growing reliance on the use of remote warfare tactics, such as partner capacity building and the use of drones. The paradox is evident: power players in the region are still interventionist, but unwilling to bear the human cost of deploying their own troops (Jazekovic 2017) and this remote presence in the region is perceived by local authorities and the population as neo-colonial. The US has not clarified its long-term strategic intentions, but both France and the EU have. The G5 Sahel Joint Force is considered a way of reducing French and foreign presence and allowing for stronger ownership of regional authorities of their own security. The stated intention is to replace Opération Barkhane and EU CSDP missions with the G5 Sahel Joint Force; however, there appears to be no timeline for when such an objective should be achieved, which leads inevitably to criticisms (RFI 2019).

But while US and European publics have taken stock of these recent developments, publicly available information within the region lags behind. The shy communication initiatives by the local government appear to be more of a result of increasing pressure on politicians to not be servient of foreign powers rather than a transparent policy choice. Journalist Ahamadou Abdoulaye Abdourahamane writes on Niamey Soir in August 2018: 'There is no independence if you are surveilled by foreign drones. We refuse this fake independence, there is no independence if our local forces cannot enter Western bases. Whatever the security threats are, military cooperation should not mean neo-colonial conquest.'[9] Journalist Seidik Abba writes 'Many

---

[9]   Loosely translated from French. Retrieved from http://www.niameysoir.com/ abdoul-ecrivain-du-sahel-lindependance-dans-la-negation-de-la-dependance-il-ny-a-

Nigeriens, such as myself, feel deep sadness for having to learn what happens in their country through the New York Times. Niger is not a federal state of the Unites States.'[10]

The real or perceived absence of a positive economic impact is another reason why the US military presence is not deemed beneficial to the region. Many inhabitants of the Tadarass neighbourhood, the closest to the Agadez US201 base, denounce the ineffectiveness of the base. Both the noise and the dust caused by the base have made US presence in Agadez hard to accept for the local residents. In addition, military presence is not even fulfilling its main purpose, which is to provide security, as foreign presence often means that the local population will more likely be targeted and become collateral damage. Moreover, there are fears that a conflict may erupt amongst regional forces and the US or French presence.

On partner capacity building, research conducted by Oxford Research Group in Mali and Kenya in September 2018 adds to this complexity by explaining how the political vacuum in capitals leads to a disarrayed coordination of troops on the ground (Knowles and Watson 2019). In Mali, Knowles and Watson (2019, 2) note 'there were a few men scattered across the multiple international military initiatives in the country run by the EU, the UN and the French without a clear sense of how these activities – in aggregate – might lead to a sustainable improvement in the capacity of their Malian partners.' In addition, HQ too often considers personnel on the ground as less relevant in the decision-making process, as the main political authority is within capital cities, which leads to a gap between those implementing strategy and those devising it. Some short-term tactics (such as preferring to train soldiers who belong to a specific ethnic group) may be quick and effective in the short term but lead to further complications in the long term in a country marred by ethnic conflict (Ibid.).

Finally, as recent research (Lyckman and Weissman 2014) shows, the use of 'light footprint war' carries several challenges which not only relate to what is mentioned above, but also to transparency and accountability. As Goldsmith and Waxman (2016, 8) point out, referring to the changes made by former President Obama, '[…] light-footprint warfare does not attract nearly the same level of congressional and especially public scrutiny as do more conventional military means.' In addition, studies on the blowback consequences of remote tactics such as drone strikes vary widely, but arguably the most complete research on such topic to date (Saeed et al. 2019) finds that 'drone strikes

---

pas-dindependance-sous-la-surveillance-des-drones-des-forces-militaires-etrangeres/

[10]   Loosely translated from French. Retrieved from https://twitter.com/abbaseidik/status/1039449240303493121?s=20

are followed by strongly elevated rates of suicide attacks' at least for the location and time period taken into consideration.[11] All such dangers of remote warfare worsen the internal problems previously mentioned above.

A role for the EU

The EU is the ideal peace broker in the region, not least because of how the region is perceived by a number of relevant member states. Lebovich (2018) argues that it is in the Sahel that some EU members believe they must fight a key battle for the future of the European project, viewing the stabilisation of the region – particularly through initiatives to hamper migration and suppress terrorist threats – as key to combating populist nationalism in their respective countries (Lebovich 2018). The EU has been intensifying its efforts in the region in response to a succession of destabilising events, from the 2012 Tuareg rebellion in northern Mali and subsequent terrorist occupation of the area to the migration crisis that moved across Europe from 2015 onwards (although European concern about the region has been increasing since 2008, if not earlier).

European leaders are also extremely proud that they saw the region as central much before other powers did and started deploying personnel very early on. The EU's main ambitions are non-military, despite having a training mission in Mali (EUTM), which means its role in the Sahel could be very different from that of member states. The EU supports several security initiatives: it has already provided €100 million to establish the African led G5 Sahel Joint Force which aims to improve security in the region and fight terrorist and criminal groups. As mentioned, the EU is itself a security player in the Sahel, with three Common Security and Defence Policy missions (EUCAP Sahel Niger, EUCAP Sahel Mali, EU training mission – EUTM – in Mali). The Council extended the mandate of the EU mission EUCAP Sahel Mali to January 2021 and allocated it a budget of almost €67 million (Council of the European Union 2019b).[12]

In addition, the EU is planning to establish a fourth CSDP mission in the region in the coming years (Lebovich 2018). It also provides more than €400 million in programmes to support stability and development in the region. For

---

[11]   The authors (Saeed et al. 2019) test whether there are elevated rates of suicide bombing activity in Pakistan during 30-day time periods immediately following drone strikes. To do so they use the Bureau of Investigative Journalism (BIJ) drone strike database, which spans from 2004 to 2017 and covers 430 strikes in Pakistan.

[12]   See European Council page on EUCAP Sahel Mali : https://www.consilium.europa. eu/en/press/press-releases/2019/02/21/eucap-sahel-mali-mission-extended-until-14-january-2021-budget-of-67-million-adopted/

example, in 2017, the EU launched a stabilisation operation in a small area of Mali, responsible for advising the Malian authorities in Mopti and Segou on governance-related issues, and supporting the planning and implementation by the Malian authorities of activities aimed at reinstating the civilian administration and basic services in the region. This team also intended to support an enhanced dialogue between the Malian authorities and the local communities (Council of the European Union 2017). However, in their drive to respond to political pressure from member states, which may be articulated in different ways, EU interventions in the region sometimes fail to adapt to conditions on the ground, potentially contributing to instability in the long run. These interventions also risk creating overly convoluted and flimsy bureaucracies both because of strategic gaps and simply because of a large presence of uncoordinated actors. The G5 Sahel force risks becoming another security architecture, which could further exacerbate the situation in the region (Schnabel 2019). As such, the EU should instead focus on a civilian rather than military component, in order to build trust with the local population and gather much needed data. The EU must also contend with member states' competing interests and overlapping missions and contributions, from France's Operation Barkhane to the recent Italian deployment – coupled with a growing US remote presence.

As mentioned, the EU is better suited to be a presence on the ground compared to other foreign forces because of local perceptions. Niger's government has recognised EUCAP Sahel Niger's value (Lebovich 2018) and gradually adapted to the mission, also increasing its participation. This shift in attitude could be seen following the onset of the European migration crisis, which showed local governments that European interest in the region was heavily dependent on the emergency and prompted demands from authoritarian regimes in the region. Elites in partner countries such as Niger show that they have learned how to use European demands to their own advantage (Koch et al 2018).

As for European remote warfare in the region and the related issue of much needed regulation changes in Brussels, the new European Defence Fund (coupled with the European Peace Facility) represents an opportunity to have a positive impact in the region. One example of this could be the acquisition and use of armed drones. Since the EU Defence Fund will not be a competence of member states[13] – such as Italy and France who are already,

---

[13]   This means that how the Fund is used will be decided upon by the European Commission (and the budget by the European Parliament) without needing to consult with member states. This is the first time in the history of the European Union that a budget for defence is an EU Commission prerogative, see my piece on the European Defence Fund: Goxho (2019), *European defence fund and European drones: mirroring US practice?*, Global Affairs

or will at some point, deploy armed drones in the region, but on EU prerogative – Brussels should focus on regulating how such missions are conducted by establishing an EU Common Policy[14] on armed drones. In this way, the EU could have a say on how such a weapon is deployed, in order not to fall for the US trap of endless remote warfare.

Moreover, the EU's integrated strategy for the Sahel centres on the idea that security, development, and governance are strongly intertwined. This does not mean that the Security-Development Nexus (whose four pillars are youth, fight against radicalisation, migration and illicit trafficking[15]) is a perfect instrument. Local civil society groups have voiced concerns around the way topics such as countering violent extremism (CVE) are treated by international actors.[16] Despite all this, though, it is undeniable that the EU strategy for the Sahel[17] presents several positive, innovative ideas for securing troubled areas, where a military approach is not deemed to be sufficient in itself to securing the region.

The European Council allowed for the establishment of a regional coordination cell (RCC) based within the European Conference on Antennas and Propagation (EUCAP) Sahel Mali (Council of the European Union 2019a). This cell includes a network of internal security and defence experts, deployed in Mali but also in EU delegations in other G5 Sahel countries. The RCC command and control structure (now renamed RACC, Regional Advisory and Coordination Cell) has recently been strengthened through an increase in the numbers of CSDP experts and moved from Bamako to Nouakchott (Ibid.). The RACC supports, through strategic advice, the G5 Sahel structures and countries, and the objective of the cell's activities will be to strengthen the G5 Sahel regional and national capacities, particularly to support the operationalisation of the G5 Sahel joint force military and police components. EUCAP Sahel Mali and EUCAP Sahel Niger will be able to conduct targeted activities of strategic advice and training in other G5 Sahel countries. The European Council envisages that in the medium to long term, the coordination hub's function will be transferred from Brussels to the structures of the G5 Sahel. The coordination hub is a mechanism which has operated under the responsibility of the EU military

---

[14]   Which could look like the one proposed by Dorsey, June 2017.

[15]   EU Special Representative Losada's interview can be viewed at: https:// africacenter.org/spotlight/eu-security-strategy-sahel-focused-security-development- nexus/

[16]   Concerns on such matters were shared during our trip to Niamey in July 2019 chiefly by the *Reseau d'Appui Aux Initiatives Locales* (RAIL) and the *Collectif des Organisations de Defense de Droit de l'Homme* (CODDH).

[17]   The European Union strategy for the Sahel can be found here: https://eeas.europa. eu/sites/eeas/files/factsheet_eu_g5_sahel_july-2019.pdf

staff since November 2017 and which provides an overview of the needs of the military G5 joint force together with the potential offers of military support from EU member states and from other donors. In other words, it is a forum which allows the matching of offers to needs.

However, in order to avoid all issues mentioned above, the EU should make sure that it establishes clear processes that would not only be beneficial to its mission, but which could also aid other foreign and regional presences. Its new focus on security and defence and its renewed interest in the Sahel are good incentives to take up more responsibility for all foreign forces operating in the region. This is clearly hard to accomplish, as security interests are not so easily negotiable, but the EU has much to offer. In order to avoid duplicating efforts, creating larger and uncooperative architectures and being perceived merely as a self-interested foreign force by the local population, the EU must ensure cooperation not just amongst its different missions in the region, but also amongst all other security actors.

In addition, it should offer a clearly demarcated and large civilian component to its missions and make sure that governance and development represent a much wider part of its agenda, starting from nudging towards a security sector reform that is more aligned with good governance and democratic principles. This is undoubtedly extremely hard as it involves negotiation and compromise with partner governments, which do not want EU interference in their internal affairs. However, given what it provides in terms of resources from development and training and its positive reputation with local communities, the EU has more leverage than it gives itself credit for and could push for best practices and positive reform.

The EU should also have in mind a clear time frame, and different and complementary objectives throughout all phases, with a particular attention to the initial and final moments. This would avoid mistakes such as the creation of other divisive community fractures, as is the case with UK forces, [18] and lack of lessons learned due to not clearly established reporting mechanisms both internally and to Brussels. Finally, the EU should have a positive communication role, not just amongst the different institutional and military actors in the region, but also with the local communities and civil society actors. The EU can be more effective compared to other actors given its

---

[18]   Knowles and Watson (2019) note in 'Improving the UK offer in Africa: Lessons from military partnerships on the continent': '[In Mali] one example is the ethnic composition of the force, which is skewed towards those from the south of the country. Accelerating the growth of an unrepresentative force in the context of ongoing conflicts between different ethnicities in Mali could be extremely detrimental to long-term security.'

connections to member states' missions, its lack of colonial and neo-colonial reputation and its resources.

Conclusion

The Sahel is experiencing a hardening of the security situation due to criminal and terrorist threats and both resources and personnel are pouring in from certain European member states, the UN and the US. Far from creating stability, this risks further exacerbating present tensions and is negatively perceived by local communities. The EU missions and EU funds could be beneficial in avoiding mistakes due to poor management and coordination amongst local and foreign forces. The EU should understand its leverage and use it to the advantage of the two key words born in the crest of the G5 security alliance: security and development.

References

Abdoulaye Abdourahamane A., 2018. 'L'indépendance dans la négation de la dépendance. Il n'y a pas d'indépendance sous la surveillance des drones des forces militaires étrangères.' *Niamey Soir*. 3 August. https://www.niameysoir. com/abdoul-ecrivain-du-sahel-lindependance-dans-la-negation-de-la-dependance-il-ny-a-pas-dindependance-sous-la-surveillance-des-drones-des-forces-militaires-etrangeres/

Africa Center for Strategic Studies. 2019. 'EU Security Strategy in Sahel Focused on Security-Development Nexus.' 7 March.

Bacchi, Umberto. 2014. 'France launches new Sahel Counter-terrorism Operation Barkhane.' *International Business Times*. 14 July.

Bailie, Craig. 2018. 'Explainer: the role of foreign military forces in Niger.' *The Conversation*. 9 September.

Barotte, Nicolas. 2018. 'French Reaper drone crashes in Niger.' *Le Figaro*. 18 November. https://www.lefigaro.fr/international/defense-les-drones-francais-armes-ont-tire-pour-la-premiere-fois-20191219

Chambas, Mohamed Ibn. 2020. 'Briefing to the Security Council on the Report of the Secretary-General on the activities of the United Nations Office for West Africa and the Sahel (UNOWAS).' United Nations Office for West Africa and the Sahel. 8 January.

Charpentreau, Clement. 2018. 'First French MQ-9 Reaper crash on record.' *Aerotime*. 19 November.

Cole, Chris. 2018. 'Could British Reaper Drones be deployed to the Sahel?' Drone Wars. 9 July.

Council of the European Union. 2017. 'Mali: European Union supports the stabilisation in the central regions of Mopti and Segou.' Press Release. 4 August. https://www.consilium.europa.eu/en/press/press-releases/2017/08/04/mali-regions-mopti-segou/

————. 2019a. 'Sahel: EU takes further steps to better support the security of the region.' Press Release. 18 February. https://www.consilium.europa.eu/en/press/press-releases/2019/02/18/sahel-eu-takes-further-steps-to-better-support-the-security-of-the-region/

————. 2019b. 'EUCAP Sahel Mali: mission extended until 14 January 2021, €67 million budget adopted.' Press Release. 21 February. https://www.consilium.europa.eu/en/press/press-releases/2019/02/21/eucap-sahel-mali-mission-extended-until-14-january-2021-budget-of-67-million-adopted/

Damon, Arwa. 2017. 'This city is a tinderbox and the US is building a drone base next door.' CNN. 21 July.

DefenceWeb. 2018. 'French Reaper crashes in Niger.' 20 November.

DefPost. 2019. 'French Air Force MQ-9 Reaper Unmanned Aircraft Crashes Near Niamey Air Base in Niger.' DP Press Briefings. 18 November.

Dorsey, Jack. 2017. 'Towards an EU Common Position on the use of armed drones.' DROI European Parliament Committee, External Policies.

European Union. 2018. 'The European Union's Partnerships with G5 Sahel Countries.' Factsheet. https://eeas.europa.eu/sites/eeas/files/factsheet_eu_g5_sahel_0.pdf

Goldsmith, Jack, and Matthew Waxman. 2016. 'The Legal Legacy of Light Footprint Warfare.' *The Washington Quarterly*, 39(2): 7–21.

Goxho, Delina. 2019. 'How can the EU improve its role in the Sahel', *VoxEurope.* 11 May.

————. 2019. 'European defence fund and European drones: mirroring US practice?' *Global Affairs,* 5(2):179–85.

Jazekovic, John. 2017. *The Use of Force in Humanitarian Intervention, Morality and Practicalities*. Routledge.

Knowles, Emily, and Abigail Watson. 2018. *Remote Warfare: Lessons Learned from Contemporary Theatres*. London: Oxford Research Group.

————. 2019. 'Improving the UK offer in Africa: lessons from military partnerships on the continent.' London: Oxford Research Group.

Koch, Anne. et al. 2018. 'Profiteers of migration? Authoritarian States in Africa and European Migration Management.' SWP Research Paper 4. Stiftung Wissenschaft und Politik. July.

Labarriere, Louis. 2019. 'Italy-Niger Agreement made public thanks to Freedom of Information Law.' *Liberties*, 15 February.

Lebovich Andrew. 2018. *Halting Ambition: EU migration and security policy in the Sahel*. European Council on Foreign Relations. 25 September.

————. 2019. 'Mapping Armed Groups in Mali and the Sahel.' European Council on Foreign Relations.

Lyckman Markus, and Mikael Weissmann. 2015. 'Global shadow war: a conceptual analysis.' *Dynamics of Asymmetric Conflict,* 8(3): 251–62.

Maclean, Ruth and Omar Hama Saley. 2018. 'Niger supresses dissent as US leads influx of foreign armies.' *The Guardian*. 14 August.

Negri, Alberto. 2018. 'Tra Francia e Italia è scontro coloniale in Niger.' *Il Manifesto*. 11 July.

Radio France Internationale. 2019. 'Mali: demonstration in Sévaré against the presence of foreign troops.' 10 October. https://rfi.fr/fr/afrique/20191009-mali-manifestation-sevare-contre-presence-troupes-etrangeres

Rogers, James. 2018. *Small States and Armed Drones*. NATO Science for Peace and Security Programme.

Saeed, Luqman. et al. 2019. *Drone Strikes and Suicide Attacks in Pakistan: An Analysis*. Action Against Armed Violence.

Schmitt, Olivier. 2018. *Allies that Count: Junior Partners in Coalition Warfare*. Georgetown University Press.

Schnabel, Simone. 2019. 'Military Cooperation in the Sahel, much to do to protect civilians.' Peace Research Institute Frankfurt (PRIF). 7 March.

Turse, Nick. 2018. 'The US is Building a Drone Base in Niger that will cost more than $280 million by 2024.' *The Intercept*. 21 August.

Vilmer, Jean-Baptiste Jeangène, and Olivier Schmitt. 2015. 'Frogs of War: Explaining the new French Military Interventionism.' *War on the Rocks*. 14 October.

VOA Africa. 2018. 'French Reaper drone crashes in Niger.' 17 November. https://www.voaafrique.com/a/un-drone-reaper-fran%C3%A7ais-s-%C3%A9crase-au-niger/4663081.html

Weiss, Caleb. 2019. 'Islamic State releases first combat video from Mali.' *FDD's Long War Journal*. 3 April.

7

The Human Cost of Remote Warfare in Yemen

BARAA SHIBAN AND CAMILLA MOLYNEUX

To governments, remote warfare is seen as an alternative way to militarily engage that is effective, cheap, and 'exceptionally precise and surgical,' with a smaller footprint and lower risk to the intervening party's troops (The White House 2013; BBC News 2012; Vitkovskaya 2012; Currier and Maass 2015). It has been argued that it facilitates the killing of 'bad guys' and 'do[es] not put...innocent men, women and children in danger' (Brennan from 2011 quoted in Purkiss and Serle 2017). These assertions, primarily made by the parties waging remote warfare, are wrong. Remote warfare shifts the burdens of warfare onto civilian populations (see Woods 2011; Purkiss and Serle 2017; Reprieve 2016, 2017, 2018; Amnesty International 2019; the Bureau on Investigative Journalism 2019; North Rhine-Westphalia Higher Administrative Court 2019). The impacts of this are evident in the case of Yemen.

The extent of civilian harm caused by drones and special forces raids in Yemen is underreported. Moreover, the reporting that has occurred has been centred mostly around civilian casualty numbers. There are significant discrepancies between the rates acknowledged by the US – criticised for its limited and superficial investigations – and those reported by NGOs, such as Amnesty International or the Bureau of Investigative Journalism.[1] The numbers of civilian casualties reported by NGOs, journalists, local populations and activists are substantially higher than those acknowledged by the US Government (North Rhine–Westphalia Higher Administrative Court 2019; the Bureau of Investigative Journalism). In 2016, the legal charity

---

[1]   See, for example, casualty recording by the Bureau of Investigative Journalism (2019), Amnesty International's investigations into civilians wrongfully categorised as terrorists by the US in post-strike data recording (2019), or reports by Reprieve, AP and Stanford University and NYU (2011).

Reprieve wrote 'the [civilian casualty] figures proposed by the US Government have been inconsistent, improbable and without even a minimal effort to provide evidentiary proof.'[2] One reason for this discrepancy is the lack of comprehensive post-strike investigations, including disregard for the findings of on the ground investigations conducted by local activists, NGOs and journalists (the Bureau of Investigative Journalism 2019; Currier and Maass 2015; Evans and Spencer 2017; Reprieve 2016). The US has not publicly addressed accounts of civilian harm beyond casualties.

Though an important endeavour, thorough monitoring of civilian casualty numbers does not capture the whole picture of civilian harm. Remote warfare operations also have significant economic, educational and mental health implications for impacted communities. Understanding these impacts of remote military operations on civilian populations is a crucial step toward evaluating the legality, legitimacy, morality and strategic significance of remote warfare.

For nearly two decades, the voices of Yemeni survivors and victims of remote warfare have been excluded from international and policy discussions on remote warfare. Drawing upon interviews with Yemenis, this chapter highlights the significant harm experienced by local populations in Yemen caused by two aspects of remote warfare: drones and special forces operations. After providing a brief background to the situation in Yemen, the first part of the chapter sets out two Navy SEAL raids as experienced by the survivors. Thereafter, the human impact of living below drones in Yemen will be examined based on interviews with 49 individuals, the majority of them from Marib. The chapter argues that the methods of harm measurement, specifically through casualty statistics only, are insufficient. Harm must also be measured in terms of the socio-economic, educational and mental health effects on local populations.

Background

Yemen has been the poorest country in the Middle East and North Africa region for some time. Today, the UN describes Yemen as 'the worst humanitarian crisis in the world,' with an estimated 80 per cent of the population – some 24 million people – at risk, and therein 14.3 million in acute need (World Bank 2019). The country is currently host to two conflicts. The

---

[2]   Since then, President Trump has further weakened Obama-era rules aimed at providing some level of operational standards and transparency (Savage 2019). For more on this, see BBC (2019) Trump revokes Obama rule on reporting drone strike deaths, 7 March. [Online] Available at https //bbc.in/2EEN7pw (accessed 14 October 2019).

civil war, which started as a conflict between the Yemeni Government and the Houthi rebels is the most widely covered – it has since developed into numerous and changing conflicts within the larger context of the civil war. A nine-state coalition largely comprised of, and led by, Saudi Arabia and the United Arab Emirates has intervened militarily to assist the Government at its request. The two Gulf states both heavily rely on US weapons and military equipment.

The second conflict is part of the covert and legally dubious US-led 'War on Terror.' Indeed, on the legal aspects of this conflict, a ground-breaking ruling by a German Higher Administrative Court in March 2019 found at least part of the programme of targeted killing by drone in Yemen unlawful (North Rhine–Westphalia Higher Administrative Court 2019; Reprieve 2019). Since 2002, Yemen has been a significant theatre for remote warfare, with the US use of drones the most widespread method for exercising lethal force (the Bureau of Investigative Journalism 2019; Evans and Spencer 2017). But it has also involved the use of special forces (Peron and Dias 2018). These counter-terrorism efforts are separate to the ongoing civil war and American support to the Saudi-led coalition. However, it is noteworthy that the Emirate's presence in Yemen has enabled it to silently assist the US War on Terror and provided US actions with increased cover, concealed by the fog of the civil war. The counterterrorism operations preceded, and will in all likelihood outlive, the civil war. The research herein pertains to the human cost of two methods of remote warfare, namely the use of drones and special forces, as part of the US-led 'war on terrorism'.

Our research

The analysis in the chapter is based primarily on interviews and conversations with a total of 49 people. Camilla Molyneux, a research consultant at the All-Party Parliamentary Group on Drones, spoke to people through semi-structured group conversations over the course of a week in Marib, Yemen in July 2018. Baraa Shiban, Case Worker at Reprieve and a youth delegate to Yemen's National Dialogue, conducted another nine phone interviews with individuals based in Bayda, Marib and Sana'a in Yemen, Amman in Jordan, and Cairo in Egypt[3] throughout 2018 and 2019. Molyneux and Shiban are both in frequent communication with the individuals interviewed.

For this chapter, survivors, victims and their families, witnesses and activists were interviewed. In order to provide a representative view of the human cost

---

[3]   At the time, some of the participants studied, had sought refuge or worked in Amman, Jordan and Cairo, Egypt.

of remote warfare by drone and special forces, the authors took care to talk to women, children and men of different ages, geographies, socio-economic status and occupations. The final section of the chapter is largely dedicated to the experiences of women. This is particularly important because existing research on the human cost of drones, due to geographical or cultural barriers, is largely based on interviews with men.

Ahead of the interviews in Marib, Molyneux consulted with a mental health expert on trauma and conflict. This informed the interview style and questions posed. It also provided several useful techniques that could be utilised in the event of an interviewee feeling anxious or experiencing flashbacks.

Finally, due to the sensitivity of this issue and fear of reprisals by the US, its allies or al-Qaeda, some names have been anonymised to protect interviewees safety. Anonymised names are marked with an asterix (*).

The Adhlan and Yakla special forces ground raids and their impacts

The raid on Yakla

On 29 January 2017, some 50 soldiers from Navy SEAL Team Six, accompanied by military dogs, carried out a raid on a small village in Yakla, Baidhaa province.[4] The aim was to capture or kill Qassim al-Rimi, the leader of al-Qaeda in the Arabian Peninsula (AQAP) (Craig 2017a). Yakla would become the second of two unsuccessful raids in the hunt for al-Rimi, that in total killed at least 66 civilians, including 31 children (Michael and Al-Zikry 2018; Reprieve 2017). The survivors of the Yakla raid refer to it as 'the night that forever changed our lives' (A. Al-Ameri and S. Al-Ameri 2019). They have further opined that 'We can never view the US the same way as we did in the past' (Ibid.).

Under the cover of a moonless sky, the SEALs approached the village without detection. Reaching the village, the SEALs lost their cover, and a two-hour-long gunfight erupted between the SEALs and tribesmen. As gunships bombarded the village, residents recall having heard the distinct buzzing sound of a drone, firing three missiles and hitting a medical unit, a school and a mosque.

The operation had been authorised by President Trump five days prior during a dinner (McFadden et al. 2017), rather than in the Situation Room which is regular procedure (Buncomb and Sharman 2017). The raid was executed less

---

[4]   See the location of the raid here https://bit.ly/2YyreB1

than a week into the Trump Administration, with Secretary of Defence James Mattis describing the operation as a 'game changer' in the campaign against AQAP (Ibid.). Within two weeks, Stephen Seche, former US Ambassador to Yemen, said the outcome 'turned out to be as bad as one can imagine it being' (Shabibi 2017). Ten children and eight women were killed, with another seven injured (Emmons 2018). Until this day, the survivors remain adamant that none of the killed villagers were AQAP members (A. Al-Ameri, A. and S. Al-Ameri 2019). Among the victims was three-month-old Asma Al-Ameri, and a heavily pregnant woman who was fatally shot in the stomach, giving birth to a baby that later died (Al-Jawfy 2017).

A widespread backlash against the US' Yemen policy followed, extending well beyond Yakla (Sanger and Schmitt 2017). This included condemnations of the raid by government officials and calls for a temporary stop to all intelligence sharing with the US (Schmitt 2017). Yemen's Foreign Minister Abdulmalik Al-Mekhlafi publicly characterised the raid as 'extrajudicial killing' (Ibid.).

Eleven-year-old Ahmed Al-Dahab was the first of many victims. Hearing something in the distance, the boy climbed onto the roof of the house. Unable to see anything, he called out 'who is there?' (A. Al-Ameri and S. Al-Ameri 2019). The question was answered by a flurry of gunshots that killed the boy. Ahmed's father, Sheikh Abdulraoof, and uncle, Sultan, met the soldiers with gunfire and were later found dead on their doorstep. An 80-year-old tribal elder, Saif Al-Jawfy, as well as eight-year-old Nawar Al-Awlaki and her mother, who was hiding inside the house with other women and children, were also killed (Ibid.).

SEAL Team Six then stormed a nearby house, according to villagers shooting indiscriminately (A. Al-Ameri and S. Al-Ameri 2019; Al-Jawfy 2017). Three children, four-year-old Aisha, five-year-old Hussein and seven-year-old Khadijah, and their father Mohammed, were all killed in this one home alone (Ibid.). Running toward his son and grandchildren's house, Abdullah Mabkhoot Al-Ameri, who survived a drone strike on his wedding day three years earlier, was shot dead in his pyjamas (Ibid.).

Witnesses said that anyone attempting to flee their homes were subject to the soldiers' fire (Ibid.). Abdullah Al-Ameri's daughter, 25-year-old Fatima, was shot and died instantaneously, whilst Fatim Al-Ameri, mother of seven, was shot in the back as she attempted to escape the soldiers, carrying her two-year-old son (Ibid.). The next morning, the boy was found alive, asleep in his dead mother's arms (Ibid.).

Amid the gunfire, a helicopter fired a missile into a third home, killing five-year-old Halima Al-Ameri (Ibid.). Yet another missile caused a roof to collapse, taking the life of three-month-old Asma Fahad Al-Ameri while she was asleep in her cot (Ibid.). Two more children, Mursil and six-year-old Khalid Al-Ameri, were killed by gunfire as they tried to escape their home (Ibid.).

In total, 23 civilians, two militants and one American soldier were killed that night (Evans and Spencer 2017). Another three US soldiers were injured when an aircraft crashed on landing – later referred to by the Pentagon as a 'hard landing' – ahead of the operation (Ibid.; McFadden et al. 2017). Upon their departure, the soldiers destroyed the aircraft to prevent it from falling into 'enemy hands.' The soldiers left behind injured and grieving survivors, destroyed homes and 120 dead sheep and goats. According to villagers this livestock which the village depended heavily on, was killed by excessive Navy SEAL fire, aimed at anything that moved (M. Al-Ameri 2017; A. Al-Ameri 2019; Reprieve 2017).

The raid on Adhlan

Despite the unsuccessful raid only two months prior, a raid on Adhlan was conducted on 23 May 2017. According to a statement by Central Command, the aim was to gather electronic equipment, including cell phones and laptops, to gain 'insight into AQAP's disposition, capabilities and intentions' (quoted in Craig 2017b).

The following section is based entirely on conversations with more than a dozen members of the Adhal family in Marib in July 2018 (Adhal 2018).

It was 1 am, Tuesday morning, and the Adhal family were fast asleep in their house, perched on the slope of a sandy hill in the Maribi desert. The air inside was warm and thick. Othman, a nine-year-old boy, was struggling to sleep and had decided to try his luck outside.

Unbeknownst to sleeping residents, some fifty US Special Forces operatives had taken up position on the nearby hillsides. Shortly thereafter, the residents were ripped out of their sleep by barking dogs and shouting. Military aircrafts thundered across the sky and descended onto the village subjecting its homes to intense gunfire. Elite soldiers charged at the houses with automatic weapons in hand, and military dogs at their feet.

During our conversation, jida,[5] gazing into the distance, stated 'the dogs were

---

[5]   The Arabic word 'Jida' translates to 'grandmother' in English.

the worst.' I don't like to talk about it, it is re-traumatising,' jida said. 'Speaking the truth about what happened that night is the only way to get some semblance of justice', a relative countered, 'Raising awareness of the effects of drones and military raids is important,' he continued. 'If the US is held accountable, this may not happen to other families.'

In one hour, the family lost three of its members. Ahead of the aerial offensive, as the SEALs were quietly taking up position on the mountains surrounding the village, Nasser, a seventy-two-year-old man, was on his way to mosque. The partially blind man mistook the SEALs for visitors and was shot dead whilst approaching the newcomers to greet them. Directly following the shooting, close by villagers stepped out of their houses, seeking dialogue with the SEALs. As a result, five men were seriously injured and another four killed: Al-Ghader, Saleh, Yasser and Shebreen. Al-Ghader was a soldier in the Yemeni military, fighting the Houthis alongside the US-supported, Saudi-led coalition in the civil war.

The family explained how machine gun fire rained down on the village and pierced into homes, causing children and women to scatter into the night. Jida ran barefoot and in her pyjamas, down the hill on which her house was located, with her grandchildren in tow. Their bare feet dug into the heavy, cold sand. From a pursuing helicopter, machine gun fire danced around the children. Fifteen-year-old Abdullah was shot dead as he attempted to escape the violence.

'It was hot,' Othman remembers, describing why he was sleeping outside. The sound of the approaching helicopters and planes scared the nine-year-old. As he ran toward the house in search of his mother, an American soldier appeared. 'I screamed,' Othman says. He was shot twice and lost consciousness. The child, who loves football and is among the brightest students in his class, rolled up his sleeves and revealed the scarring from two gunshot wounds on his forearms. His eyes were wide; the pain from that night was painted across his face. An hour after the SEALs' arrival, the shooting subsided. Othman, who had regained consciousness, found his mother weeping, cradling the dead bodies of his two older brothers.

The survivors struggle with a multitude of visible and invisible scars. Two of the Adhal brothers will live out the rest of their lives with disabilities; one had his left leg amputated at the knee, the other brother is in Cairo for treatment at a monthly cost exceeding the family's yearly savings. An elder pulled out a picture of the young man in a hospital bed, connected to multiple life-saving machines. In the photo, bandages are tightly wrapped around his head, legs and arms. A woman, tears streaming down her face, witnessed her husband die in their bed, beside her. She was one of many survivors experiencing overwhelming, traumatic flashbacks at home. 'The only room she'll set foot in is the kitchen,' a family member said quietly.

The raid has had a significant impact on the Adhal children's mental health, causing them to suffer from insomnia, depression and anxiety. One boy, the family said, was near unrecognisable. Following his father's death, the once outgoing child who loved school was silent, worried and refused to leave the house. 'We have to force him to go to school' the family commented. Following the raid, his little sister was forced to grow up fast. At only about 10 years old, she was already stepping in as a substitute adult. In a neighbouring house, a young boy lost his hearing. Only being able to communicate with his family through self-made hand gestures, the child has had to process his experiences alone.

There are visible signs of the raid everywhere. Walls are littered with bullet holes, all the windows have been replaced, and only a few meters from the home, a blown-up pickup sits on its head. 'I hate this house', jida said with tears running down her face. 'Every evening I walk ten minutes into the desert to sleep under a tree. I cannot bear to sleep here anymore.'[6]

The impact of drones on local populations

US military operations in Yemen, including intelligence, surveillance and reconnaissance (ISR), are aimed at disrupting terrorist activities and preventing attacks by AQAP and the Islamic State in Yemen. Under President Obama, drones became a central counterterrorism tool. His Presidency marked a policy shift from Bush-era large-scale, boots-on-the-ground operations, extraordinary rendition and 'personality' strikes, to 'signature'[7] strikes, with new deployments small in scale, often conducted by special forces or working through, with and by partners.

The following sections are primarily based on accounts collected from across Marib and will briefly outline the effects of the US drone programme on

---

[6]   Following the Navy SEAL raid, Reprieve, a London-based human rights organisation, found the family was mistakenly targeted. *The Intercept* and *the Bureau of Investigative Journalism* both reported that five civilians were killed. Despite comprehensive evidence, the US continues to dispute claims of civilian casualties.

[7]   'Between 2002 and 2007, the Bush administration reportedly focused targeted killings on "personality' strikes targeting named, allegedly high-value leaders of armed, non-state groups' (Stanford University and NYU 2012 12; Hudson et al. 2011). President Obama expanded the programme to include 'far more...'signature' strikes based on a 'pattern of life' analysis' (Ibid.; Cloud 2010; Klaidman 2012).

civilian communities in the province. As previous studies largely have relied on testimonies from men,[8] women and children are the primary focus herein.

The buzzing sound of drones

The overwhelming impact caused by drones on civilian populations was a constant theme throughout the interviews. Responding to a frequently voiced statement in US and UK military and policy-making circles, namely that drones can neither be seen nor heard from the ground in Yemen, a young boy was adamant to counter this narrative, describing detailed visuals (Women 2 2018). NGOs and reporters – local or with partners on the ground – maintain that drones can be heard. The authors of this chapter have both heard drones in Yemen. In fact, this issue would have been laid to rest some time ago, if the voices of people with first-hand experience – the voices emphasised herein – were afforded the same legitimacy and audience as those dominating the international conversation on drones.

A man living elsewhere said that 'the drones fly lower during the night, they are so loud it can be difficult to hear the TV!' (Adhal 2018). Moreover, several families reported that under President Trump, the aircrafts fly at a lower altitude and with more frequency (Adhal 2018). What has caused this change of behaviour remains speculation; however, Trump's relaxation of safeguards might be one explanation.[9]

Impacts on mental health

> Every day, we kiss our loved ones goodbye, not knowing if we will ever see them again. It is like living in a constant nightmare from which we cannot wake up. (Faisal bin Ali Jaber, 2015)

In March 2019, a German High Administrative Court found that drones have a significant impact on civilians living below them (North Rhine-Westphalia Higher Administrative Court 2019). This echoes reports from almost every individual interviewed, stating that frequent and unpredictable drone appearances caused widespread and constant fear, frustration and apathy (Abia* 2018; Badia* 2018; Cala* 2018; Adhal 2018). The inability to predict who, when or where a strike would hit, as well as the failure to keep their

---

[8]   Reports by Stanford University and NYU (2012) and by Williams (2015) have documented the impact of drones on civilians in Pakistan and Afghanistan, respectively.
[9]   For more on this, see BBC (2019) 'Trump revokes Obama rule on reporting drone strike deaths. 7 March. https //bbc.in/2EEN7pw (accessed 14 October 2019).

family safe, caused a lot of grievance, especially among heads of households (Ibid.; Women 1 2018; Women 2 2018). Noting the open-ended rhetoric surrounding counterterrorism drone operations in Yemen, many caretakers expressed immense frustration and anguish at not being able to provide a safe future for their children (Ibid.). Upon the loss of her son to a drone strike, one woman expressed the difficulty in providing for her grandchildren and coping with her own sorrow, 'I cry for my son every day' (Cala* 2018).

Responding to questions about their own experiences, mothers frequently reverted to the health of their children. Among the interviewees, it was commonly believed that the young population experienced the most severe mental health effects (Abia* 2018; Adhal 2018; Badia* 2018; Cala* 2018; Women 1 2018). The mothers reported that children living in proximity to drone- and air strikes have experienced significant mental health problems, including insomnia, depression, mood swings, anxiety, apathy and fear (Ibid.). Every time a drone is heard or seen, one mother explained, the whole village of 1900 people evacuate into the desert in their cars. They only return once the drone has left the area. This procedure is repeated without failure at every sighting or audio confirmation of a drone, on average two to four times a month. Abia*, a mother from Marib, (2018) said 'When they [the children] hear the drones, they run home from school calling for their mothers, then everyone gets into their cars and evacuate the village.' Abia* (2018) added that a number of the children would scream 'the Americans are coming to kill us!' Another woman described the rapidly deteriorating mental health of her son:

> My son has attempted to commit suicide multiple times. He walks to a nearby busy road and lies down [...] He says he wants to join his father [who died from a drone strike] in heaven. (Baida*, a mother from Marib, 2015).

Pregnant women are another group frequently identified as particularly vulnerable, with numerous reports of women losing their unborn children (Al-Qadhi 2018; Sabaa 2018). One activist[10] reported that the blast wave caused a nearby woman to miscarry (Al-Qadhi 2018). Miscarriages, local women say, have been caused by physical injury from drone strikes and military raids and, they believe, immense psychological stress experienced during these operations (Ibid.; Sabaa 2018; Women 2 2018).

Accounts also described a link between the presence of drones, and people applying self-restrictions on their movements: 'One day I was about to go

---

[10]   Entessar Al-Qadi is a Maribi activist. She was a delegate to the National Dialogue, (March 2013 to January 2014) , representing Marib Governorate and women.

outside, but my husband stopped me. He could hear the drone. I stayed inside all day,' Al-Qadi explained (2018). Similarly, a mother described how, following a drone strike only meters from her family's tent in an informal camp for internally displaced people, she immediately imposed restrictions on her children's movements, banning them from playing outside (Derifa* 2018). In areas with frequent drone presences, children are left without safe places to play. These accounts echo previous research from Pakistan and Afghanistan (Stanford University and NYU: 2013; Williams 2015). Researchers from Stanford University and NYU (2012, 55), for example, found

> [T]hose interviewed stated that the fear of strikes undermines people's sense of safety to such an extent that it has at times affected their willingness to engage in a wide variety of activities, including social gatherings, educational and economic opportunities, funerals and that fear has also undermined general community trust.

Impacts on education

According to interviewees, the impacts of drones on education are linked to those on mental health (Abia* 2018; Adhal 2018; Badia* 2018; Cala* 2018). Through conversations with parents and children, it became clear that some children could access school, whereas others stayed at home due to the perceived or actual danger travelling in areas frequented by drones. One mother reported that she kept her children home as a means to protect them after a drone struck only meters from their tent in an unofficial IDP camp (Derifa* 2018). A group of mothers, most of them primary breadwinners after their husbands left to fight in the civil war, described how desperately they wanted their children to get an education and a better life (Women 1 2018; Women 2 2018). It was particularly important that their daughters go to school, so that if left in a position similar to their mothers, they would not have to resort to begging (Ibid.). For some families, a working child was of financial necessity, following the death by drone of a breadwinning family member. Nevertheless, the women feared that those who left school early would miss out on crucial education that might be essential to secure future employment. This was a significant worry as the pressure on the job market in Marib skyrocketed, following a population-increase from 40,000 to 2 million in only five years (Government of Marib 2018).

Impacts on family finances

Drone strikes have direct impacts on survivors, notwithstanding the obvious impacts caused by the loss of a loved one. In some instances, wives and even children, of deceased breadwinners are forced to work in order to compensate for lost income. In a society where women primarily conduct unpaid labour, and where paid work has only recently opened up to women (and is highly competitive), several of the interviewees said they resort to begging on the streets of Marib, or from their neighbours (Women 1 2018).

For the Al-Adhal family, the loss and injury of the family's three primary breadwinners – two were left disabled and one killed – had a detrimental financial effect. The two surviving men have, due to their injuries, been discharged from their positions in the Yemeni national military without financial support. One lost his leg, and the other is receiving medical treatment in Cairo, Egypt. 'The medical treatment is more than US$600 per month. Each month, we have about 25 Saudi Rial [the equivalent of US$6.7] to spare,' said an elder (Al-Adhal 2018). The remaining US$593, the family have to borrow.

Property, including homes, vehicles, tools and livestock have been damaged or destroyed in air strikes and raids (A. Al-Ameri 2019; Sabaa 2018). In a country where the UN estimates that 80 percent of the population – 24.1 million people – are at risk of hunger and disease (World Bank 2019), the overwhelming majority of survivors cannot afford to rebuild or replace damaged property. As such, the damage caused by strikes has led some survivors to permanently lose important income, take on debt and/or go hungry (A. Al-Ameri 2019; Al-Adhal 2018; Al-Qadhi 2018). Some survivors are forced to continue living in damaged or makeshift homes that are constant reminders of their traumatic experiences (Al-Adhal 2018). As mentioned, in Yakla, the raid left homes and vehicles destroyed and livestock dead. Similarly, the Al-Adhal's home remains riddled with bullets and, more than two years on, a missile-damaged pickup truck rests on its head in the sand only meters from the home.

Lack of accountability and failure to investigate and acknowledge civilian harm

Of the 329 US air strikes and counterterrorism operations conducted in Yemen, killing an estimated 207 to 325 civilians (the Bureau of Investigative Journalism 2019), not a single apology has been issued or compensation offered. A number of family members and survivors have reached out to the Central Command (CENTCOM), the US military Command overseeing activities in the Middle East and beyond, seeking an explanation or apology. An organisation that has assisted survivors with this correspondence says CENTCOM has never responded. A survivor from the Yakla raid said he 'thought an investigation would be launched, and that we would receive an apology and compensation' (S. Al-Ameri 2019). Instead, he described 'the behaviour of the US was aggressive and unaccountable' (Ibid.).

Survivors explain how the lack of recognition amounts to their lives and innocence – and those of their killed relatives – being disregarded. In a letter to Presidents Obama and Hadi (of Yemen), Faisal bin Ali Jaber (2019) wrote that 'neither the current US or Yemeni administrations bother to distinguish between friend and foe.' The lack of distinction was also raised by Zabnallah Saif Al-Ameri. The Yakla survivor, who lost nine members of his extended family, including five children, said

> It is true they were targeting al-Qaeda but why did they have to kill children and women and elderly people? If such slaughter happened in their country, there would be a lot of shouting about human rights. When our children are killed, they are quiet. (Shabibi 2017)

Thoughts for the future

Civilian harm extends beyond casualties

The voices elevated in this chapter show the importance of expanding our understanding of civilian harm beyond that of casualties only, to include all harm, whether it be indirect, physical and/or psychological. For example, the consistent psychological pressures of a seemingly imminent attack, which in turn, considerably lowers quality of life, was emphasised by a significant number of interviewees. The account of harm set out in this chapter requires further research. This must be understood by military and political decision makers and inform policy development and decision-making processes pertaining to the use of military force.

Informed decision-making requires complete knowledge of all effects of military operations

The lack of understanding of the comprehensive civilian harm caused by counterterrorism operations in Yemen is reflected in policy- and decision-making circles in Washington and across the capitals of US allies, where civilian accounts are unrepresented. This has significant unacknowledged consequences, some of which are highlighted in this chapter. Without complete knowledge of the effects of drone and special forces operations – including the human cost and the number of civilian casualties and injuries – the US is unable to make fully informed decisions about its counterterrorism operations. Crucially, full information must include the experience of ordinary people, women and children.

Civilian harm is bad for strategy

The lack of understanding of, or ambivalence toward, all aspects of civilian harm can be counterproductive. Civilian harm is recognised by journalists, academics and military leaders to run counter to overall military strategic aims (Gul 2017; Jaffer 2016; BBC 2010; Petraeus and Kolenda 2016; Reprieve 2016; Stanford University and NYU 2012; Scahill 2016; Waldman 2018). Whereas protecting civilians has taken a more central role in conventional warfare,[11] it has not translated well to the battlefield of remote warfare, where militaries consistently fail to consider the relation between strategic effectiveness and extensive civilian harm. In a 2013 letter to President Obama and Yemen's President Hadi, Faisal bin Ali Jaber wrote

> With respect, you cannot continue to behave as if innocent deaths like those in my family are irrelevant. If the Yemeni and American Presidents refuse to engage with overwhelming popular sentiment in Yemen, you will defeat your own counterterrorism aims. (quoted in Kutty 2014)

Our research suggests links between civilian harm and strategic effectiveness may extend to remote warfare. Anecdotal evidence, such as children believing the 'Americans are coming to kill us,' or the belief that innocent deaths are treated as 'irrelevant,' suggest the US might be losing the human ground and requires further examination. Moreover, it calls for a re-evaluation of the current policy and research into more proportionate counterterrorism methods.

Local populations as allies

Local populations can, and should be, recognised as potential resources and allies. More research needs to be conducted into alternative, long-term and peace-focused 'security' partnerships. Marib would be a natural place to trial this. During our interviews, Maribis went to great lengths to vocalise their opposition to AQAP (Adhal 2018; Sho'lan 2019; Governor Al-Aradah 2019). Moreover, systematic efforts to counter terrorism at the provincial, tribal and

---

[11]   See Petraeus (2010) and Petraeus and Kolenda (2016) to read more about the strategic importance of protecting civilian casualties in conventional war; and the Iraq Inquiry (2016) to read about the legal and moral obligations to protect civilian populations.

grassroot level have been in practice since 2002.[12] More recently, communities have requested photographs of terrorists and hotlines to directly communicate with security forces:

> If they provide us with pictures, we can let the government know when we see them (AQAP) [...] This way the government can make sure other innocent people do not get hurt (Al-Adhal 2018).[13]

At the provincial level, Governor Al-Aradah and Chief of Security Sho'lan have both expressed the need for increased local control and ownership over counterterrorism operations (2018, 2019). This, Sho'lan said, will improve the public's perception of the local Government as accountable and legitimate (2019). The Governor (2019) concurred and added 'it is more efficient to tackle extremists with local forces.' In conversations with US and UK officials, he has emphasised this point, requesting that local forces replace US drones and special forces raids. This move would place counterterrorism operations within Marib's conventional security structure, with the aim of capturing, prosecuting and trying suspects in the court of law (Abdul* 2018).

To conclude, this chapter has set out the widespread harm caused by drones and special forces operations in Yemen, particularly how it extends far beyond casualties only. It has highlighted the need to protect civilians from all harm, ranging from physical to psychological. Further research aimed at understanding the full spectre of civilian harm and how to mitigate it in policy and decision-making is of the utmost importance. With the aim of identifying an effective, sustainable and long-term solution, such research would benefit from a peace-making lens as well as widespread local participation. The methodologies of the ground interviews and group conversations and inclusivity, particularly that of people with few economic means, women and children, presented in this chapter provide a good starting-point.

References

Primary sources, interviews conducted by Baraa Shiban and Camilla Molyneux

---

[12] For instance, tribes have enforced a strict policy of excluding and reporting (to the Government) any tribal member that joined a terrorist network.

[13] Because a large number of US drone strikes in particular appear to be signature strikes - targeting unidentified individuals primarily based on metadata - the US would be unable to provide civilian populations with pictures of their targets.

Abdul (anonymised senior official). 2018. *Interview with a senior official in Marib.*

Abia (anonymised). 2018. *Interview with a woman (and mother) from rural Marib.*

Al-Ameri, Aziz. 2019. *Interview with Aziz Al-Ameri.*

Al-Ameri, Saleh. 2019. *Interview with Saleh Al-Ameri.*

Al-Ameri, Mohsina. 2017. *Interview with Mohsina Al-Ameri.*

Al-Aradah, Sultan. 2019. *Interview with the Governor of Marib, Sheikh Sultan Al-Aradah.*

Al-Arafat, Sultan. 2019. *Interview with Sultan Al-Arafat.*

Al-Jawfy, S. 2017. *Interview with Al-Jawfy.*

Al-Jawfy, Sheikh. 2019. *Interview with Sheikh Al-Jawfy.*

Al-Qadhi, Entessar. 2018. *Interview with Entessar Al-Qadhi, a prominent human rights activist and National Dialogue Representative from Marib.*

Al-Adhal. 2018. *Conversation with 12 Adhal family members and neighbours at the Adhal home in rural Marib.*

Badia (anonymised). 2018. *Interview with a woman and mother from rural Marib.*

Cala (anonymised). 2018. *Interview with a woman and mother from rural Marib.*

Derifa (anonymised). 2018. *Interview with a woman and mother living with her family in an unofficial camp for internally displaced people outside of Marib.*

Government of Marib. 2018. *Interview with local Maribi bureaucrats about Marib Governorate.*

Sabaa (anonymised). 2018. *Interview with local student and activist.*

Sho'olan, Abdulghani. 2019. *Interview with Marib's Chief of Security, Sho'lan.*

Women 1. 2018. *Conversation with a group of 10 women living in rural Marib.*

Women 2. 2018. *Conversation with a group of 7 women living in rural Marib.*

Secondary sources

Amnesty International. 2018. 'European Assistance to Deadly US Drone Strikes.' 19 April. Accessed 14 October 2019. https //www.amnesty.org/en/latest/news/2018/04/european-assistance-to-deadly-us-drone-strikes/

Amnesty International. 2019. 'US military shows appalling disregard for civilians killed in Somalia air strike.' 30 September. Accessed 14 October 2019.

The Bureau of Investigative Journalism. 2019. Drone Strikes in Yemen. Accessed 14. Oct 2019. https //www.thebureauinvestigates.com/projects/drone-war/yemen

BBC News. 2010. 'Petraeus focuses on civilians in Afghan directive.'4 August. Accessed 28 July 2019. https //www.bbc.co.uk/news/world-south-asia-10869738

————.. 2012. 'Obama defends drone strikes in Pakistan.' 31 January. https://www.bbc.co.uk/news/world-us-canada-16804247

Becker, Jo, and Scott Shane. 2012. 'Secret 'Kill List' Proves a Test of Obama's Principles and Will.' *New York Times.* 29 May. Accessed 14 October 2019. http //www.nytimes.com/2012/05/29/world/obamas-leadership-in-war-on-alqaeda.html?pagewanted=all

Buncomb, Andrew and Jon Sharman. 2017. 'Donald Trump not in Situation Room for "botched" Yemen raid that killed up to 30 civilians and one US Navy SEAL.' *The Independent.* 3 February. Accessed 14 October 2019. https //www.independent.co.uk/news/world/americas/donald-trump-not-in-situation-room-yemen-raid-30-civilians-killed-us-navy-seal-dead-first-military-a7561596.html

Cloud, David, S., 2010. 'CIA Drones Have Broader List of Targets.' *LA Times.* 5 May. Accessed 14 October 2019. http //articles.latimes.com/2010/may/05/world/la-fg-drone-targets-20100506

Craig, Iona. 2017a. 'Death in Al Ghayil: Women and Children in Yemeni Village Recall Horror of Trump's "Highly Successful" SEAL Raid.' *The Intercept*. 9 March. Accessed 28 July 2019. https //theintercept. com/2017/03/09/women-and-children-in-yemeni-village-recall-horror-of-trumps-highly-successful-seal-raid/

Craig, Iona. 2017b. 'Villagers say Yemen child was shot as he tried to flee Navy Seal raid.' *The Intercept*. 28 May. Accessed 14 October 2019. https // theintercept.com/2017/05/28/villagers-say-yemeni-child-was-shot-as-he-tried-to-flee-navy-seal-raid/

DeYoung, Kevin. 2012. 'After Obama's remarks on drones, White House rebuffs security questions.' *The Washington Post*. 31 January. Accessed 27 July 2019. https //www.washingtonpost.com/world/national-security/after-obamas-remarks-on-drones-white-house-rebuffs-security-questions/2012/01/31/gIQA9s2LgQ_story.html?utm_term=.23e27d9dc538

Emmons, Alex. 2018. 'Pentagon says 35 killed in Trump's first Yemen raid - More than twice as many as previously reported.' *The Intercept*. 20 December. Accessed 27 July 2019. https //theintercept.com/2018/12/20/yemen-raid-investigation/

Evans, Michael. and Richard Spencer. 2019. 'President's first Navy Seal raid was doomed from the start.' *The Times*. 3 February. Accessed 28 July 2019. https //www.thetimes.co.uk/article/trump-s-first-navy-seal-raid-was-doomed-from-the-start-lhwnm67cl

Foa, Maya. 2018. 'Trump's secret assassinations programme.' Reprieve. Accessed 27 July 2019. https //reprieve.org.uk/update/game-changer-trumps-new-attacks-on-human-rights/

Gul, Ayaz. 2017. 'Pakistan Army Chief Slams US Drone Operation.' *VOA News*. 14 June. Accessed 27 Jul. 2019. https //www.voanews.com/east-asia-pacific/pakistan-army-chief-slams-us-drone-operation

Hudson, Leila, Colin S. Owens and Matt Flannes. M. 2011. 'Drone Warfare Blowback from the New American Way of War.' *Middle East Policy,* (Fall). Accessed 10 October 2019. http //www.mepc.org/journal/middle-east-policy-archives/drone-warfareblowback-new-american-way-war

Jaber, bin Ali Faisal. 2014. 'Letter to Obama and Hadi on Yemeni drones.' *Middle East Monitor*. 4 February. Accessed 28 July 2019. https //www.middleeastmonitor.com/20140204-letter-to-obama-and-hadi-on-yemeni-drones/

————. 2015. 'Declaration of Faisal bin Ali Jaber.' ECCHR. Accessed 27 July 2019. https //www.ecchr.eu/fileadmin/Juristische_Dokumente/Hearing_Statement_FaisalBinAliJaber_engl.pdf

Jaffer, Jameel. 2016. *The Drone Memos. 1st ed*. New York: The New Press.

Klaidman, Daniel. 2012. 'Drones How Obama Learned to Kill' (excerpt from Klaidman's book KILL OR CAPTURE THE WAR ON TERROR AND THE SOUL OF THE OBAMA PRESIDENCY, infra note 53). *Daily Beast.* 28 May. Accessed 14 October 2019. https://www.newsweek.com/drones-silent-killers-64909

Kutty, Faisal. 2014. 'The drone "blowback": Drones fuel "blowback" and undermine core principles of American identity.' *Al Jazeera*. 18 July.

McFadden, Cynthia, William M. Arkin and Tim Uehlinger. 2017. 'How the Trump Team's First Military Raid in Yemen Went Wrong.' *NBC News*. 1 October. Accessed 14 October 2019. https //www.nbcnews.com/news/us-news/how-trump-team-s-first-military-raid-went-wrong-n806246

Michael, Maggie, and Maad Al-Zikry. 2018. 'The hidden toll of American drones in Yemen Civilian deaths.' *Associated Press*. 14 November. Accessed 27 July 2019. https //www.apnews.com/9051691c8f8a449e8bb6fd684f100863

North Rhine-Westphalia Higher Administrative Court. 2019. 'Judgement from 19/3/2019 - 4 A 1361/15 - Wording of the oral pronouncement of the judgement.' *Faisal bin Ali JAber and others v. the Federal Republic of Germany*. Unofficial translation prepared for the European Center for Constitutional and Human Rights (ECCHR), one of the NGOs supporting the claimants in this case.

Office of the Press Secretary, The White House. 2013. 'Remarks by the President at the National Defense University.' 23 May. Accessed 27 July 2019. https //obamawhitehouse.archives.gov/the-press-office/2013/05/23/remarks-president-national-defense-university?redirect=TDM/ObamaSpeech

Perry, Bruce, and Maia Szalavit. 2017. *The boy who was raised as a dog And Other Stories from a Child Psychiatrist's Notebook What Traumatized Children Can Teach Us about Loss, Love and Healing*. 3rd edition. Basic Books.

Peron, Alcides Eduardo dos Reis, and Rafael de Brito Dias. 2018. '"No Boots on the Ground": Reflections on the US Drone Campaign through Virtuous War and STS Theories." *Contexto Internacional,* 40(1): 53–71.

Petraeus, David, and Chris Kolenda, C., 2016. 'We learned through experience the importance of preventing civilian casualties in today's wars.' *Defense One*. 7 July. Accessed 28 July 2019. https //www.defenseone.com/ideas/2016/07/obama-asked-military-plan-protect-civilians-heres-one/129681/

Purkiss, Jessica, and Jack Serle. 2017. 'Obama's Covert Drone War in Numbers Ten Times More Strikes Than Bush.' *The Bureau of Investigative Journalism.* 17 January. Accessed 15 October 2019. https //www.thebureauinvestigates.com/stories/2017-01-17/obamas-covert-drone-war-in-numbers-ten-times-more-strikes-than-bush

Report of a Committee of Privy Counsellors. 2016. *The Report of the Iraq Inquiry*. London: House of Commons: 170–217. Accessed 28 July 2019. https //webarchive.nationalarchives.gov.uk/20171123122743/http //www.iraqinquiry.org.uk/the-report/

Reprieve. 2016. *Opaque Transparency: the Obama Administration and Its Opaque Transparency on Civilians Killed in Drone Strikes*. London: Reprieve. Accessed 27 July 2019. https //reprieve.org.uk/wp-content/uploads/2016/06/Obama-Drones-transparency-FINAL.pdf

————. 2017. *Game Changer An investigation by Reprieve into President Donald Trump's secret assassination programme and the massacre of Yemeni civilians in the villages of Yakla and Al Jubah*. London: Reprieve. Accessed 27 July 2019. https //reprieve.org.uk/wp-content/uploads/2017/10/2017_10_31_PRIV-Yemen-Report-UK-Version-FINAL-FOR-USE.pdf

————. 2019. 'UK "on notice" after court rules Germany failed in duty to protect innocent civilians from US drones.' 19 March. Accessed 14 October 2019. https //reprieve.org.uk/press/uk-on-notice-after-court-rules-germany-failed-in-duty-to-protect-innocent-civilians-from-us-drones/

————. N/A. 'The 5 most important things that happened in the first year of Trump's illegal drone war.' Accessed 14 October 2019. https //reprieve.org.uk/update/5-important-things-happened-first-year-trumps-illegal-drone-war/

Sanger, Daniel, and Eric Schmitt. 2017. 'Yemen Withdraws Permission for US Antiterror Ground Missions.' *The New York Times*. 7 February. Accessed 29 July 2019. https //www.nytimes.com/2017/02/07/world/middleeast/yemen-special-operations-missions.html

Savage, Charlie. 2019. 'Trump Revokes Obama-Era Rule on Disclosing Civilian Casualties From US air strikes Outside War Zones.' *The New York Times*. 6 March. Accessed 28 July 2019. https //www.nytimes.com/2019/03/06/us/politics/trump-civilian-casualties-rule-revoked.html

Scahill, Jeremy. 2016. *The assassination complex: inside the government's secret drone program*. 1st edition. Simon and Schuster.

Schmitt, Eric. 2017. 'Women Killed in Yemen Raid Were Qaeda Fighters, Pentagon Says.' *The New York Times*. 30 January. Accessed 27 July 2019. https //www.nytimes.com/2017/01/30/world/middleeast/yemen-raid-women-qaeda.html

Shabibi, Namir. 2017. 'Nine young children killed: The full details of botched US raid in Yemen.' *The Bureau of Investigative Journalism.* 9 February. Accessed 27 July 2019.https //www.thebureauinvestigates.com/stories/2017-02-09/nine-young-children-killed-the-full-details-of-botched-us-raid-in-yemen

Stanford University and NYU. 2012. *Living Under Drones Death, Injury, and Trauma to Civilians From US Drone Practices in Pakistan.* International Human Rights and Conflict Resolution Clinic at Stanford Law School and NYU Global Justice Clinic at NYU School of Law.

Vitkovskaya, Julie. 2012. '9 revealing statement Obama about transparency and drone strike', *Washington Post*. 30 January. Accessed 14 October 2019. https //wapo.st/29bryRT

Waldman, Thomas. 2018. 'Vicarious warfare: The counterproductive consequences of modern American military practice.' *Contemporary Security Policy*, 39(2): 181–205, DOI 10.1080/13523260.2017.1393201

Williams, John. 2015. 'Distant intimacy space, drones, and just war.' *Ethics and international affairs*, 29(01): 93– 110. Accessed 14 October 2019. http // dro.dur.ac.uk/14486/1/14486.pdf?DDD35+DDC68+dpl0jcw+d700tmt

Woods, Chris. 2011. 'US claims of "no civilian deaths" are untrue.' *The Bureau of Investigative Journalism.* Accessed 14 October 2019. https //www.thebureauinvestigates.com/stories/2011-07-18/us-claims-of-no-civilian-deaths-are-untrue

World Bank. 2019. The World Bank in Yemen. Accessed 14 Oct. 2019. https //www.worldbank.org/en/country/yemen/overview

8

Human Rights and Civilian Harm in Security Cooperation: A Framework of Analysis

DANIEL MAHANTY

On 13 March 2019, the US Senate passed a resolution intended to bring an end to US support for the Saudi-led military campaign in Yemen. For nearly four years, the US had supported its ally with equipment, intelligence and logistical support amidst a growing record of large-scale civilian casualty incidents and in the wake of a mounting human toll from the conflict. But with each report or photo of the devastation wrought by the Saudi-led campaign, and the murder of *Washington Post* journalist Jamal Khashoggi, cracks began to emerge in the long-held assumption that the strategic partnership was immune to American public concerns about the Kingdom's human rights record. While the Senate's vote on Yemen is one of the more recent, and the most notorious cases, it is one of many circumstances in which American strategic interests have been compromised by insufficient attention given to the risks of civilian harm – and to including civilian casualties, human rights abuses and humanitarian crises either created by, or associated with, its security cooperation and security assistance activities.

A note on terminology: While security cooperation and security assistance have different definitions in sources of US Government guidance, this paper will use the term 'security cooperation' to include security assistance activities for purposes of simplicity.

The case of support to Saudi demonstrates one way in which the United States Government may encounter a policy obstacle, in the form of public scrutiny, that results from harmful policies and practices of a partner over which it has little or no control. But civilian harm and human rights violations committed by America's partners can also undermine progress toward the security interests that security cooperation is intended to achieve in more fundamental ways. For example, human rights abuses by a partner can counteract joint efforts to counter terrorism by eroding the public's trust in the legitimacy of the partner state (and by extension, the United States), or even by increasing the number of the disaffected who may turn to violence in response to state-sponsored abuse (UNDP 2017, 5, 80). Meanwhile, providing security assistance in fragile states, where such harms may be more likely or prevalent, can also aggravate conflict, undermine stabilization, or counteract peacebuilding efforts (Kleinfeld 2018, 282–285). Finally, human rights abuse or civilian harm may be symptomatic of larger institutional deficits or structural phenomena (e.g. corruption) that greatly diminish the likelihood that security cooperation will ever achieve its desired outcomes.

Having a sounder understanding of risks specific to human rights and civilian harm could, at minimum, allow the US government to optimise the desired returns on security cooperation with fewer attendant costs. This chapter sets out a summary of a framework developed by Center for Civilians in Conflict, by which the US – and even other states or institutions – might assess the human rights and civilian harm risks involved with undertaking security cooperation activities with particular partners. It also presents a selection of factors to consider in the design of the partnership to ensure 'interoperable', i.e. compatible, approaches to risk mitigation. As such, the US Government might use the framework in one of three ways: 1) *ex ante* policy analysis of its partners to help formulate security cooperation strategies; 2) ongoing analysis of risks that can be used to head off significant problems arising from harm or to monitor progress or improvements; and 3) a means of designing and customizing risk mitigation measures specific to, and appropriate for, partnerships of vastly different character.

Scope of application

For purposes of this analysis, security cooperation can include a variety of activities along a spectrum of involvement, from partnered operations involving both countries in the military or security operations, to the provision of advice, operational or logistical support, arms sales, and training and education (Knowles and Watson 2018, 3). While partnership with civilian law enforcement may be implicated by this framework, and certain analytical criteria may apply to police activities, the analysis is primarily geared toward application to cooperation between, and with, military forces.

Meanwhile, the framework is based on the hypothesis that security cooperation can correspond directly or indirectly with human rights or civilian harm (which may or may not be *per se* lawful) in one or more ways, *inter alia:*

1. Assistance may directly enable harm caused independently by the partner (e.g. missiles sold by the US are used to strike hospitals, or a unit trained by the US commits a mass atrocity)
2. Assistance, provided over time, may more generally enhance the lethality of a partner force in a vacuum of appropriate constraints
3. Assistance may distort, or exacerbate, the distortion of defence or security priorities (thereby depriving civilians of needed security services)
4. Support or assistance may abet harm as a direct result of jointly conceived or planned activities or operations (e.g. the US acts on intelligence from a partner that yields a targeting error and civilian casualties)
5. Assistance or materiel provided in the absence of adequate controls may be diverted to other groups or individuals who subsequently harm civilians
6. Support or assistance may occur independently, but incidental to significant forms of harm (the US sells major military equipment to a country with an abusive police force)
7. Support to security forces with a record of human rights abuses or civilian harm in the absence of accountability risks tolerating or even endorsing impunity
8. Acts of security cooperation may elevate the tolerance threshold of risk for human rights abuses or civilian harm (e.g. the US pressures a partner to act more assertively in a military operation)

Limitations on current approaches to risk management

The United States government does, by all accounts, take human rights and civilian harm risks into consideration as it plans and implements its security cooperation activities. However, while the constituent elements of its current approach to risk management are each independently worthwhile, they may not *together* constitute an approach that is optimally designed for both anticipating and managing risks. Restrictions on assistance based on past conduct, and especially conduct related to human rights violations (such as through the US 'Leahy Laws'[1]) can be effective for reducing material support to specific units who have already committed violations. But examining the risk of partnership on the basis of past conduct may not adequately consider variables that relate to the likelihood of harm in the *future*. Meanwhile, leaning on legal analysis to ensure an act of support is technically 'lawful' under its own interpretation of obligations under state responsibility may have little bearing on the practical effects of harms experienced by civilians and may not comport with public or even international perceptions of lawfulness or legitimacy (Shiel and Mahanty 2017, 9).

---

[1] For background on the Leahy Laws, see: 'Leahy Law' Human Rights Provisions and Security Assistance: Issue Overview. https://fas.org/sgp/crs/row/R43361.pdf

Moreover, technical assessments of competencies or capability gaps, may not adequately consider legal, political, or other environmental factors that may have a significant bearing on the likelihood of human rights abuses or civilian harm, and in so doing, may ignore the interaction between variables at the political and operational – or even tactical – levels. Although certain institutional variables may proxy for other kinds of risks (e.g. inadequate legal safeguards may represent operational risk of human rights violations) and are therefore worthy of attention, attending to the risk of partnership with a focus on one category of analysis at the expense of others can lead to narrowly conceived solutions which fail to address the source of risk. Finally, attempts to reduce risk through purely technical interventions like training, or the provision of less than lethal equipment or precision weapons may yield some positive benefits under certain conditions, but may not overcome institutional deficiencies or overcome overriding political incentives that enable harm to occur (Dalton et al. 2018, 23).

In diagnosing risks, and in designing measures to mitigate them, an overly narrow scope of application can carry unintended consequences. First, policymakers may over-estimate the extent to which any one safeguard actually mitigates the most pressing risks (e.g. the government may defend its support to a partner based on its confidence in the capacity for human rights training to overcome gaps in accountability). Second, the government may overlook opportunities to achieve the intended outcome of a security cooperation activity while also managing the risk through more proactive, and effective controls. Finally, any of these approaches may fail to account for the way a cooperative relationship, and the context in which it exists, can change over time.

A framework of analysis for civilian harm and human rights risks in security cooperation

To establish a more suitable approach to managing the different kinds of risks associated with different kinds of security cooperation, policymakers, program managers and those who evaluate government practices from the outside, might look toward a new framework of analysis that establishes the basis for designing programs with adequate controls and safeguards in place. A well-designed framework would 1) align with the emerging policy consensus about the necessary components of effective and sustainable security cooperation programs; 2) identify the variables within several categories of analysis (e.g. political, legal, and operational) that correspond with the risks most worthy of attention; 3) allow for customised application to a specific security cooperation activity; 4) consider the risks that derive both from the 'partner' as well as from the 'acts of partnership', and 5) lend itself to relatively clear policy prescriptions that flow from the diagnosis it presents.

Each category of risks in the framework that follows includes a sample of representative risk indicators, along with an associated claim about how any perceived or deal deficits may impair the government's ability to achieve a desired end-state with its partner. Ideally, the government could use the framework to make policy decisions about the viability of partnership (e.g. as prerequisite conditions); to design and modify partnership activities; and to prioritise and integrate risk mitigation measures. Among options that may be considered in response to analysis conducted on the basis of this framework, policymakers or program managers might consider moderating or reducing any forms of assistance that materially enhance the lethality of security forces (Kleinfeld 2018, 283–285).   The government may also elect to sequence support or establish pre-requisite requirements (e.g. demonstrating a particular competency) prior to providing certain kinds of assistance or partaking in certain kinds of partnership activities; (Shiel and Mahanty 2017, 32), or to require more direct oversight ('operational end-use monitoring') (Lewis 2019, 28) of operations that benefit from support or assistance.

1. Political factors

The political context in which security partnerships take place has direct and inextricable bearing on the risk of human rights or civilian harm in at least three discernible ways. First, security relationships between states are shaped by, and bound to, the alignment or misalignment of political intentions and objectives of each state (Tankel 2018, 308–310). When objectives are misaligned, a state risks participating in security operations that do not serve its core interests, but also partially 'owns' any negative consequences (Rand 2018). As importantly, the political organisation of the state and the political orientation of its security forces carry significant implications for the ways in which the state employs the use of violence (e.g. in the violation of human rights). When a state such as the United States enhances the capacity of state security institutions, they are better able to serve whatever political purpose they ultimately serve (Kleinfeld 2018, 284). If state security forces have been instrumentalised to preserve the economic or political power of a rent-seeking elite, then the security forces may in turn instrumentalise the use of force in service of that objective. In these cases, enhancing the capacity of the state may contribute to the state's hold on the monopoly on the use of violence, but in service of illegitimate ends.

The extent to which the political character of security institutions should and can affect the assessment of human rights and civilian protection risks largely depends on the type of cooperation activity and the scale of divergence between the status quo and a desired end state. Indicators in this category are most useful during policy deliberations about whether, and how much, the

United States should pursue its foreign policy aims through security cooperation activities.

Figure 1: Political/Strategic (partner State).

Variable	Desired End-State	Risk	Sample Assessment Question(s)
Political Character of Security Services	Security institutions are apolitical and free of corruption.	Security institutions use violence and extortion to preserve political status quo rather than to provide for public safety and security.	What is the relationship between security institutions and the political apparatus of state? What evidence exists of corruption in the security sector?
Political Commitment	Demonstrated commitment to human rights and civilian harm mitigation by the partner state.	Absence of political commitment to human rights or minimizing harm leads to civilian harm.	What evidence exists of a political commitment to protecting rights and minimizing harm?
Civilians in Strategy	Policies and plans reflect emphasis on civilian harm mitigation.	Absence of emphasis on minimizing harm to civilians in strategic planning leads to inadequate or inconsistent protection.	How does civilian harm mitigation feature in strategic plans?
Factionalism or Bias in Security Institutions	Security institutions are apolitical and representative.	Politicized or factionalized institutions may lead to institutional divisions or even abuse of specific groups.	What evidence exists that security institutions are apolitical and free of factionalism or bias?

Figure 2: Political/Strategic compatibility

Variable	Desired End-State	Risk	Sample Assessment Question(s)
Shared Threat Perceptions	Partners Share common understanding, views of threat	Variation in threat perception leads to risk of misappropriation or diversion of support.	How does the US understand the threat? How does the partner understand the threat?
Political and Strategic Objectives	Partners have similar or mutually reinforcing strategic and political objectives.	Variance in strategic or political objectives leads to misallocation of support (partner wants support for x reason, US wants to support for y reason).	What are the strategic objectives advanced by the partnership? Do the partners agree?
Alignment on end-state	The U.S. and its partners share common perspectives of end-state.	Variance of perspective on end-state distorts views of how human rights and civilian harm mitigation relates to outcomes and desired end-states.	What does success look like? (Political, Security, Military) What is the exit strategy for the US?
Means/resources	Partners agree on scope and nature of means to achieve the political and strategic objectives.	Disagreement on means leads to 'downstream' incompatibilities in partnership with potentially negative repercussions.	What are the means that the US and the partner are ready to/ seek to apply?

2. Security governance factors

The proclivity of security forces to engage in human rights abuse or conduct that results in harm may also depend on the extent to which security institutions are subject to appropriate constraints in the form of internal controls, external oversight[2] and accountability measures (DCAF 2018, 7–10). While the methods and means of exercising oversight and accountability of security forces may vary depending on the circumstance, they should work together to ensure the prevention of and accountability for human rights violations and should work to rectify shortcomings in security governance. Partnering with, or supporting, a state's security forces that operate in a vacuum of accountability and oversight increases the risk that the assistance could directly lead to abuse or harm. There is also the risk that the providing state will become associated with harm to which it has made no explicit contribution (Goodman and Arabia 2018, 33–35; Shiel and Mahanty 2017, 31).

While it is important to evaluate the institutional makeup of the partner forces and the oversight bodies that govern them, it is equally important to assess the extent to which the partnership itself is subject to oversight and accountability. For example, it may be appropriate or prudent to subject certain features of a partnership to parliamentary oversight to ensure a degree of 'informed public consent', even if such arrangements are not bound by treaties requiring parliamentary approval. Indicators within this category are most useful during policy deliberations, but also in determinations of the kinds of cooperation the US may be willing to undertake – or should avoid, such as lethal assistance – given any major institutional deficits.

---

[2]   According to DCAF, 'Oversight is a comprehensive term that refers to several processes, including ex-ante scrutiny, ongoing monitoring, and ex-post review, as well as evaluation and investigation. Oversight of security services is undertaken by a number of external actors, including the judiciary, parliament, National Human Rights Institutions (NHRI) and ombuds institutions, National Preventive Mechanisms (NPM), audit institutions, specialised oversight bodies, media and NGOs. Oversight should be distinguished from control as the latter term implies the power to direct an organisation's policies and activities. As such, control is typically associated with the executive branch of government.' (DCAF, 6)

Figure 3: Security Governance – accountability and oversight

Variable	Desired End-State	Risk	Sample Assessment Question(s)
Investigations (Military/ Administrative)	Partner state forces have process for conducting internal investigations into reportable incidents.	Lack of process for administrative investigations prevents accountability or suppression of wrongful acts.	Does the Partner state have a process for conducting administrative investigations? Is the process effective, as a function of timely, impartial, and thorough?
Legislative Oversight	Security activities are subject to legislative oversight.	Absence of, or inadequate, public oversight may lead to abuse or gaps in accountability.	What role does the legislature play in overseeing security policies and activities?
Independent Oversight	Security forces and their activities are monitored by official but independent oversight bodies.	Absence of independent oversight bodies may lead to security force abuse.	What independent oversight bodies monitor the activities of security forces?
Internal Accountability	Partner nation security forces have effective processes, to include personnel and regulations, for internal accountability.	Absence of effective internal accountability mechanisms may lead to security force abuse or impunity for abuse.	What is the structure and role of internal accountability mechanisms within the partner security force?
Civilian harm or human rights complaints / Allegations	Partner nation has process for receiving complaints or allegations of civilian harm or human rights violations from civilians or third parties (NGOs).	Absence of complaints mechanism prevents appropriate response.	How do civilians and other external actors, such as media, IOs and NGOs make reports of human rights violations or civilian casualties to partner nation government? How are complaints managed?

Figure 4: Security Governance interoperability

Variable	Desired End-State	Risk	Sample Assessment Question(s)
US support to accountability institutions	US provides technical assistance or support to security governance, accountability, or oversight mechanisms.	Lack of support to institutions may aggravate risk of abuse or impunity.	What forms of support does the US provide to security governance, accountability, or oversight?
Investigations – Interoperability	US and Partner state are able to conduct joint investigations	Lack of protocols for joint investigations, to include chain of custody for evidence, interviewing witnesses, and public reporting may lead to inadequate investigations.	What protocols exist for joint investigations? Can and does the US support partner state nation in the conduct of investigations? Do the US and partner have the ability to share evidence?
CIVCAS Complaints – Interoperability	US and Partner state have a process for communicating about reports of civilian casualties	Communication gaps in civilian casualties reporting prevents appropriate response to harm.	How do the US and Partner state communicate when reports of civilian casualties are received?

3. Legal compliance and interoperability factors

Questions related to the interaction between security cooperation and international law fall loosely into two broad and distinct, but related, categories. The first category includes questions that relate to the obligations each state has under international law, how the cooperation activity accrues responsibility to each for their own action and how each might share responsibility for the actions of the other. The degree of legal liability imparted to states that participate in security cooperation activities largely depends on the applicable legal theory of state responsibility (and how it is being interpreted), and may depend on the nature and significance of the action, and whether and how much a state 'knows' about any internationally wrongful acts it is supporting. [3] This category of

---

[3]   See Articles on Responsibility of States for Internationally Wrongful Acts, ILC Yearbook 2001/II (2)(ARSIWA or Articles on State Responsibility), Articles 6, 16, 17, and 47. http://legal.un.org/ilc/publications/yearbooks/english/ ilc_2001_v2_p2.pdf.

concerns also raises questions about any gaps that are created with respect to attribution and accountability for actions among states in 'coalitions of the willing' (Tondini 2017, 11–13). Common Article 1 of the 1949 Geneva Conventions also obliges states to 'undertake to respect and to ensure respect' for international humanitarian law (IHL), which could, in theory, create an additional obligation on states to promote respect for IHL through both positive and negative means available.

Figure 5: Legal (partner state)

Variable	Desired End-State	Risk	Sample Assessment Question(s)
Clear and Appropriate Understanding of Applicable Legal Frameworks	Partner clearly and accurately acknowledges status of conflict and applies correct body of law.	Mis-identification of applicable legal framework may affect lawfulness of use of force, detentions.	How does the partner state understand the legal status of the conflict?
Consistent interpretation of legal obligations and concepts	Interpretation of key concepts and obligations under international law are consistent and appropriate.	Divergent or inconsistent interpretation of legal obligations or key concepts may expose civilians to risk of harm.	How does the partner state understand and interpret key legal concepts, such as participation in hostilities?
IHL Compliance	The partner has a history of compliance with IHL principles of the use of force, to include adequate precautions taken in attack, proportionality, and distinction.	IHL compliance history by partner is disputed or flawed.	What accounts has the US received from respectable internal and/or external sources on its partner's respect towards IHL, e.g. ICRC?

Other questions relate to the degree of 'legal interoperability'[4] between the partnering states that ensures compatible interpretation and application of the law (Goddard 2017, 211, 212). Legal Interoperability might also include a consideration for how any formal or informal legal arrangements dilute or strengthen accountability to the law, e.g. in the form of status of forces agreements that relieve one of the states of accountability (Hussein, Moorehead and Horowitz

---

[4]   Defined by the International Committee of the Red Cross as 'a way of managing legal differences between coalition partners with a view to rendering the conduct of multinational operations as effective as possible, while respecting the relevant applicable law.' International Committee of the Red Cross, International Humanitarian Law and the Challenges of Contemporary Armed Conflicts: 32–33.

2018). Indicators in this category may be most useful for circumstances in which the US is providing operational support or is involved with the partner in joint operations (so-called 'partnered operations').

Figure 6: Legal interoperability

Variable	Desired End-State	Risk	Sample Assessment Question(s)
Status of Forces Agreement (SoFA)	SoFA is legal and lawful, and includes ability for civilians to levy complaints or make claims for loss/ damage.	SoFA prevents accountability for wrongful acts, or prevents civilians from seeking civil damages.	What does the SoFA cover? Does the SoFA or other bilateral agreements in any way prevent accountability?
Comparative legal obligations	Each partner understands legal obligations created by specific role (e.g. state responsibilities are clear).	Misidentification of positive and negative obligations enables partner abuse or negligence.	How does the US understand its legal obligations under theories? of state responsibility? How does the US understand its obligations to "ensure respect" for IHL?
Liability – Effective Control	Any Partner state forces under effective control of US forces act in accordance with international obligations.	Conduct of forces, to include any internationally wrongful act, may be attributed to US forces if in effective control.	Are there any circumstances where US forces are in "effective control" of partner units?

4. Operational factors

Operational capacity is unlikely to overcome factors at the political level that enable, encourage or tolerate human rights abuses. But some degree of tradecraft competency is necessary for translating a political commitment to protecting rights or preventing harm into operational reality. In the case of law enforcement operations, police, or security forces acting as police, can avoid or reduce harm through tactical proficiency in escalation of force or the effective use of non-lethal methods (OHCHR 2004, 27). Similarly, military forces can take meaningful steps toward preventing, mitigating, and responding to any ham they cause in the course of their own operations during the conduct of hostilities (Lewis 2019, 38). (Note: the small selection of sample variables here applies to military operations and civilian harm, rather than law enforcement activities and human rights. They are meant to be representative of the kinds of variables that may be appropriate.)

Figure 7: Operational

Variable	Desired End-State	Risk	Sample Assessment Question(s)
Tactics and Procedures	Doctrine, to include formal guidance, manuals, instructions, and code of conduct includes and emphasizes preventing and responding to civilian harm.	Weak or inconsistent inclusion of civilians in doctrine may lead to gaps or risks.	How do civilians configure in sources of partner doctrine?
Command Culture	Commanders and military leadership emphasize importance of preventing and responding to civilian harm.	Absence of command culture that emphasizes civilian harm, or creates incentives to de-emphasize civilian protection may enable harm.	What evidence exists in the host nation of a command climate that emphasizes civilian harm mitigation?
Roles and Responsibilities	Operational components of the military have dedicated assets for managing civilian harm issues.	Lack of dedicated assets with specialized knowledge and capabilities leads to inconsistent and unpredictable methods of preventing and responding to civilian harm.	Do operational components and personnel have clear guidance and understanding of their roles relative to civilian harm mitigation?
Weapons	Weapons, armaments, and explosives are deployed with adequate precautions.	Lack of understanding of weapons and munitions secondary and tertiary effects increase risks of civilian harm.	Which weapons delivery systems are available and used by partner state forces? Are forces appropriately trained on them?

Continued.

Figure 7: continued

Variable	Desired End-State	Risk	Sample Assessment Question(s)
Targeting	Systems and processes for dynamic and deliberate targeting are adequate and calibrated for minimizing the risk of civilian harm.	Poor or inadequate targeting systems and processes increase the risk of civilian harm.	What are the targeting procedures (including fires clearance processes and command authority) for artillery and air strikes? Do all partner units calling for surface to surface and air to surface fires include trained forward observers and forward air controllers? Do they support with positive identification and assessments of civilian harm? Does the partner state employ a no-strike list?
Post-Operational Assessment (e.g. BDA)	Partner state has a process for assessing the effects of operations.	Lack of assessment process increases risk that civilian harm will be missed or under-estimated.	Does the Partner state have a process for conducting post-operational analysis, such as strike logs and BDA (that include information on civilian harm)?
Pattern of Life Analysis and Collateral Damage Estimates	Partner state has a process for accurately analyzing pattern of life and collateral damage estimates that leads to compliance with IHL and to minimizing civilian harm.	Lack of process for accurate analysis of pattern of life or collateral damage may lead to civilian harm.	Does the Partner state have an effective process for analyzing patterns of life? What is the Partner state process for collateral damage estimates?

Similar to the other categories within this framework, operational risks derive not only from the characteristics of the partners, but also from the acts of cooperation they undertake. Just as risks can emerge from variations in the understanding and application of law, so too can they stem from gaps in interoperability in joint or partnered security operations, to include the ways in which partners plan, prepare, and execute missions, and in the ways in which they respond to allegations or reports of harm or abuse (Dalton et al. 2018, 20). When providing materiel or logistical support, the US may benefit from ensuring it is able to undertake 'operational oversight' (or 'operational end-use monitoring') so that it can better understand the nature of the military activities it is supporting. (Lewis 2019, 28).

Figure 8: Operational interoperability.

Variable	Desired End-State	Risk	Sample Assessment Question(s)
Intelligence Collection	Intelligence sharing is vetted for bias and errors.	Biased and outdated intelligence may lead to mis-allocation of resources or positive identification error.	Does the U.S. have processes in place to control for bias in intelligence from the partner state? How do political biases of the partner state military affect the risk of intelligence error?
Post-Operational Assessment (Interoperability)	Partner state and US have process for comparing results of post-operational analysis (e.g. after action reviews and BDA).	Lack of coordination in assessing effects of operations leads to gaps in internal reporting and records that can substantiate or refute external reports.	Do the US and Partner state have a process for evaluating the effects of operations together, e.g. BDA and after-action?
Joint Concept of Operations (CONOPs)	Partners and US forces operate on the basis of the same or complementary concept of operations.	Variation in concept of operations may reflect divergent objectives, threat perceptions, or emphasis on civilian harm mitigation.	Do partner state forces operate with same CONOPs? In what ways to CONOPs complement each other or diverge?

5. Civil-military relations and civil society factors

Finally, the risk of civilian harm or human rights issues can be managed through constructive and meaningful interaction between security forces and civilians, to include the consideration of civilian perspectives in the design and implementation of security cooperation activities. Two broader categories of engagement are particularly germane to this framework. The first is the capacity of either, or both, partners to engage in effective civil-military relations in the context of security operations, which includes having channels available for understanding and responding to community protection needs, to include the risks faced by civilians; and the degree to which the military and the public have a shared understanding of 'security.' Important to the analysis are the ways in which security forces communicate with civilians in advance of, during and after operations and the extent to which the military avoids creating or transferring risks to civilian populations through certain activities. These can include seeking intelligence, maintaining an active presence near civilians, or failing to consider safety, security, or privacy of the communities or individuals with whom they engage.

The second category of criteria that can be pertinent to the design of security cooperation is the role and functionality of civil society as both a source of public oversight of security force activities, and a potential signal of informed public consent to partnership arrangements (Dalton et al. 2018, 12). Governments undertaking security cooperation activities should be wary of tacitly encouraging or abiding by cases in which the partner government has placed explicit, or implicit, restrictions on civil society and its ability to act as a meaningful check on government activities. In some cases, civil society, for political reasons or otherwise, may not be present, active, or coordinated to engage government on security policy. In these cases, the United States should ensure that it seeks meaningful and voluntary input from sources that can serve as a proxy – or it should calibrate its activities to account for the additional risk imposed by the lack of civil society as a meaningful check.

Conclusion

Even as the US Government shifts its stated defence priorities toward 'great power competition', and reduces the numbers of its own forces from theatres such as Afghanistan and Iraq, the era of security partnership, and the emphasis on 'by, with, and through' is likely here to stay. If anything, as great powers seek to avoid the devastation of direct confrontation by competing through more indirect means, the range and scale of security cooperation activities and security assistance may actually increase. In the absence of the proper controls, these activities may run the risk of distorting local systems of governance, exacerbating corruption, or contributing to human rights abuse.

The mere availability of a more organised set of risk indicators will not automatically generate the appropriate policy response now, or in the future, until policymakers recognise the importance of taking human rights and civilian harm more seriously when designing and implementing security cooperation and assistance activities. After all, the US government was not oblivious to the human rights record of its Saudi partners, nor unable to diagnose gaps in its interest or capacity to avoid civilian harm using US-manufactured weapons. Even so, a more comprehensive framework of analysis may assist policymakers and security cooperation program managers in anticipating problems before they arise, compelling clear policy choices, and in identifying the areas where mitigation measures are best placed throughout the life cycle of a partnership.

Figure 9: Civil-military and civil society.

Variable	Desired End-State	Risk	Sample Assessment Question(s)
Civilian-Military (Civ-Mil) dialogue on protection issues	Voluntary forum exists for discussing protection concerns, reporting incidents, and providing rationale for security force operations.	Absence of dedicated voluntary channel for communicating concerns leads to gaps in civ-mil dialogue about civilian protection concerns. What forms of civ-mil dialogue exist?	Are civilians able to express concerns about protection directly or indirectly to security forces? Are consultations done in such a manner that they avoid increasing risk to civilians?
Communication with Civilians	Partner state forces communicate clear expectations and guidance to civilians in order to minimize harm.	Lack of clear instruction or guidance places civilians at greater risk of harm as a result of miscommunication.	How does partner state communicate expectations and instructions to civilians? Do partner forces understand motivations or limitations that civilians face in complying with guidance?
Civ-Mil Deconfliction	Two-way channel of communication exists to deconflict humanitarian movement and access.	Absence of channel of communication increases risk of harm to humanitarian workers or civilians.	Does the partner state have a process for deconflicting movement and access by humanitarian actors?
Safety and Security	Partner state takes steps to ensure security, safety, and privacy for any participant in civ-mil dialogue.	Failure to consider security of civilians may increase exposure of safety and security risk to civilians.	What protocols exist to ensure the privacy, safety, and security of civilians who choose to engage the military in civ-mil dialogue?
Freedom of Association	Civil society and individuals are free raise to exercise freedoms of association and expression, to include on issues related to civilian harm and security forces.	Restrictions on civil society leads to inadequate oversight and accountability of security forces, lack of informed public consent for security policies.	What restrictions does civil society face in openly raising security policies?

Figure 10: Civil-Military interoperability.

Variable	Desired End-State	Risk	Sample Assessment Question(s)
Coordination with humanitarian	Protection Actors U.S. and partner forces respect humanitarian actors' reliance upon their neutrality, impartiality and operational independence. Partner and U.S. forces are cognizant of and act with a clear distinction between the role and function of humanitarian actors and of political and military actors in mind.	Failure to systematically coordinate with and respect humanitarian principles of humanitarian protection actors in shared theatre leads to less effective civilian harm prevention, response and mitigation.	How do partner state and U.S. forces interact and coordinate activities with humanitarian protection actors? What are communication channels with humanitarian actors prior and during partnered operations? How do partner and U.S. forces proactively avoid 'blurring the lines' between their
Risk Exposure	Partner state and US forces take steps to avoid exposing civilians to harm or risk of by association.	Failure to consider and control risks of association leads to greater risk exposure for civilians	How do U.S. and partner state forces ensure that security operations (including intelligence collection) do not expose civilians to greater risk of harm?
Consultation with civil society on protection issues	US government consults actively with any willing organization or person who has informed perspectives on civilian harm, human rights prior to and during partnership activities.	Failure to consult with civil society or individuals may lead to incomplete picture of partner force or protection concerns and lack of information among the public about purpose and nature of US activities.	Does the US actively consult with civil society to get a variety of perspectives? Does the US consult with civil society to inform them about the nature and purpose of US security cooperation?

References

Beittel, June S., Lauren Ploch Blanchard, and Liana Rosen. 2014. *"Leahy Law" Human Rights Provisions and Security Assistance: Issue Overview.* Congressional Research Service.

Dalton, Melissa, Shannon Green, Hijab Shah and Rebecca Hughes. 2018. *Oversight and Accountability in US Security Sector Assistance: Seeking Return on Investments.* Center for Strategic and International Studies. Washington, DC.

Dalton, Melssa, Jenny McEvoy, Hijab Shah and Daniel Mahanty. 2018. *The Protection of Civilians in US Partnered Operations.* Policy, Center for Strategic and International Studies. Washington, DC.

DCAF. 2018. *International Standards and Good Practices in the Governance and Oversight of Security Services.* Policy, Geneva: Geneva Center for Democratic Control of the Armed Forces (DCAF).

Goddard, David. 2017. 'Understanding the Challenge of Legal Interoperability in Coalition Operations.' *Journal of National Security Law and Policy,* 9: 211–232.

Goodman, Colby, and Christina Arabia. 2018. *Corruption in the Defense Sector: Identifying Key Risks to US Counterterror Aid.* Policy, Washington, DC: Center for International Policy.

Hussein, Rahma, Alex Moorehead and Jonathan Horowitz.'Niger Facing Pressure to Ensure U.S. and French Drone Strikes Comply with Human Rights Law.' *Just Security.* 6 September.

International Committee of the Red Cross. 2015. *International Humanitarian Law and the Challenges of Contemporary Armed Conflicts.*

Kleinfeld, Rachel. 2018. *A Savage Order.* New York: Pantheon.

Knowles, Emily, and Abigail Watson. 2018. *Remote Warfare: Lessons Learned from Contemporary Theatres.* Research, London: Oxford Research Group.

Lewis, Larry. 2019. *Promoting Civilian Protection during Security Assistance: Learning from Yemen. Policy* Washington, DC: Center for Naval Analysis.

OHCHR. 2004. 'Human Rights Standards and Practice for the Police.' Training, Geneva: Office of the United Nations High Commissioner for Human Rights.

Rand, Dafna. 2018. 'Extricating the United States from Yemen: Lessons on the Strategic Perils of Partnered Operations.' *Lawfare*. 25 November. Accessed 1 June 2019. https://www.lawfareblog.com/extricating-united-states-yemen-lessons-strategic-perils-partnered-operations

Shiel, Annie, and Daniel Mahanty. 2017. *With Great Power: Modifying US Arms Sales to Reduce CIvilian Harm*. Washington, DC: Stimson Center and Center for Civilians in Conflict.

Tankel, Stephen. 2018. *With Us and Against Us*. New York: Columbia University Press.

Task Force on Extremism in Fragile States. 2019. *Final Report of the Task Force on Extremism in Fragile States*. Policy, Washington, DC: United States Institute of Peace.

Tondini, Matteo. 2017. 'Shared Responsibility in Coalitions of the Willing.' In *The Practice of Shared Responsibility in International Law*, edited by André Nollkaemper and Ilias Plakokefalos: 701–732. Cambridge: Cambridge University Press.

UNDP. 2017. *Journey to Extremism in Africa*. Research. New York: United Nations Development Program, Regional Bureau for Africa.

9

Security Cooperation as Remote Warfare: The US in the Horn of Africa

RUBRICK BIEGON AND TOM WATTS

Speaking in 2007, US Defense Secretary Robert Gates argued that the 'most important military component in the War on Terror is not the fighting we do ourselves, but how well we enable and empower our partners to defend and govern themselves' (Gates 2007).[1] Consistent with this claim, the Bush, Obama and Trump administrations have all engaged in a variety of efforts to build the capacity of foreign security forces to address security-related threats. This has required the Department of Defense (DOD) to develop a broad spectrum of bilateral and multilateral military activities under the rubric of security cooperation. These activities, of which the more widely debated security force assistance is a subset,[2] have been a critical component of contemporary US foreign and counterterrorism policy (Biddle, Macdonald and Baker 2017; Stokes and Waterman 2017; Tankel 2018b). They are also integral to the debates on remote warfare (Watson and Knowles 2019; Watts and Biegon 2017, 2019). Security cooperation is defined by the Pentagon as all

---

[1]   The authors would like to thank Maria Ryan, Simone Papale and the reviewers for their comments on earlier versions of this chapter. Any mistakes remain our own.

[2]   As explained in the Joint Publication 3–20 security cooperation, security force assistance 'is the set of DOD [security cooperation] activities that contribute to unified action by the [United States Government] to support the development of the capacity and capabilities of [Foreign Security Forces] and their supporting institutions, whether of a [Partner Nation] or an international organisation (e.g., regional security organisation), in support of US objectives' (Joint Chiefs of Staff 2017, vii).

> [...] interactions, programmes, and activities with foreign security forces (FSF) and their institutions to build relationships that help promote US interests; enable partner nations (PNs) to provide the US access to territory, infrastructure, information, and resources; and/or to build and apply their capacity and capabilities consistent with US defense objectives. (Joint Chiefs of Staff 2017, v)

This chapter introduces security cooperation as a tool of remote warfare, both in a general sense and in the specific case of US counterterrorism operations in the Horn of Africa. We argue that there is a twin security/strategic logic to its use: it functions to build the capacity of foreign security forces to deny terrorist organisations safe havens within their own borders or region; and to help secure American access to bases, airspace and foreign security personnel, 'thicken' political partnerships with overseas governments and to create new patterns of cooperation, influence and leverage.[3]

The notion that security cooperation is 'political' is not novel. It underpins much of the recent practitioner-oriented literature on the limits of recent Western partner capacity building efforts (Biddle, Macdonald and Baker 2017; Matisek 2018; Reno 2018; Tankel 2018b). A greater focus on the politics animating the use of security cooperation activities rather than the politics of the agents receiving this assistance, however, provides an alternative calculus for revisiting the debates on their effectiveness. Much of the existing academic-practitioner dialogue on US security cooperation activities in the Horn of Africa has focused on the failures to build capacity in Somali and regional security agents (Reno 2018; Ross 2018; Williams 2019). When the political dimensions of US military assistance are discussed, it is usually within the context of how misalignments in the political interests of the US and recipient have undermined the efficacy of partner building efforts. We argue that this is problematic because a greater sensitivity to the *twin* security and strategic logics of security cooperation can potentially help us better understand the apparent puzzle of why these activities have persisted despite their well-documented military failures.

To be clear, we are not arguing that it is only the strategic logics of security cooperation which explain their use, nor are we arguing that the various forms of access their use generates offsets the failure to build partner capacity. Moreover, we are sensitive to the methodological challenges of documenting the relationship between security cooperation and securing the different forms

---

[3]   Beyond this, security cooperation activities in Africa can also be theorised as having a political-economy component, see (Stokes and Waterman 2017, 838–840; Ryan, 2020).

of access discussed above, aware that there is not necessarily a clear 'transmission belt' between the two. Nonetheless, as we document through engagement with various primary source material, reference to the Cold War era use of military assistance, and the empirical study of contemporary US counterterrorism operations in the Horn of Africa, the use of security cooperation as a tool of remote warfare can be understood to have supported the pursuit of wider strategic goals.

Our analysis unfolds in three stages. Section 1 introduces the major trends in post-war military assistance with a particular focus on the Bush, Obama and Trump presidencies. Section 2 unpacks the twin security and strategic logics of security cooperation as an instrument of remote warfare. This framework is used in the final section of this chapter to examine the role of security cooperation in US counterterrorism operations in the Horn of Africa. Somalia, our principal case study, has been the centre of American security co-operation activities in Africa during the last decade (Ross 2018). It is also emblematic of the US' support for 'Fabergé egg militaries' which are 'expensive, shiny, and easy to break' (Matisek 2018, 278–279). Whilst their use has been greater in Somalia than in military operations elsewhere in Africa, this case is recognised to be representative of the wider demand for, and use of, security cooperation in fragile states (Reno 2018, 498).

Security cooperation in US foreign policy: From the Cold War to Trump

Military assistance, of which security cooperation is one component, has long been a key tool of American foreign policy.[4] The US is estimated to have provided military assistance to over 100 states after 1945 (Kuzmarov 2017). During the Cold War, an estimated \$390 billion was spent on military and developmental assistance (Matisek 2018, 273). This served multiple strategic purposes. Beyond helping partners defend against communist expansion, it was a key conduit through which the US stabilised access to overseas markets (Kolko 1988) and helped secure access to overseas bases (Kuzmarov 2017).

As Defence Secretary Robert McNamara told Congress during the 1960s, the US provided military aid because 'military officers were the coming leaders of their nations. It is beyond price to the United States to make friends with such men' (House of Representatives 1963, 291). Military assistance, he empha-sised, generated 'important economic by-products for our foreign policy with respect to the stability and economic progress of the less developed and emerging nations' and helped secure 'access to overseas bases and

---

4   For a more detailed discussion of the relationship between Security cooperation and the other channels of US military assistance, see (White 2014).

installations' (House of Representatives 1963, 60). All three of these dynamics were apparent in in the Horn of Africa. Prior to the communist coup which overthrew Emperor Haile Selassie in 1974, Ethiopia had received $286 million worth of military aid following the Second World War (Kuzmarov 2017). Thereafter, as the patterns of material support were reordered to reflect the region's new political landscape, the flow of military assistance was redirected toward neighbouring Somalia (Oberdorfer 1977). As the *Washington Post* candidly reported at the time, the US agreed to 'provide $40 million in weapons in return for the use of Somali air bases and ports' (J. Ross 1981).

Such practices continued after the Cold War. Military assistance has been integral to post-9/11 efforts to deny transnational terrorist organisations safe havens in fragile states (Biddle et al. 2017; Ryan 2019; Tankel 2018b; Watts and Biegon 2017). Billions of dollars were spent by the US and its coalition partners training, equipping and advising tens of thousands of Afghani and Iraqi soldiers as part of the counterinsurgency campaigns in both countries. Whilst these activities have been smaller in scale, military assistance has also been central to what Maria Ryan has coined the 'War on Terror on the periphery' (Ryan 2019, 2020). Key for counterterrorism operations in Africa were the shifts laid out in the 2006 Quadrennial Defence Review, a highly influential defence planning document which distilled the Pentagon's evolving approach to irregular warfare (Ryan 2019, 144–152). It outlined a number of important adjustments to US defence strategy, including a shift from larger-scale military interventions toward fighting 'multiple irregular, asymmetric operations' (DOD 2006, vii). This required:

> Maintaining a long-term, low-visibility presence in many areas of the world where [US] forces do not traditionally operate. Building and leveraging partner capacity will also be an absolutely essential part of this approach, and the employment of surrogates will be a necessary method for achieving many goals. Working indirectly with and through others, and thereby denying popular support to the enemy, will help to transform the character of the conflict. (DOD 2006, 23)

These commitments remained a core component of Obama's counterterrorism policies. The 2012 Defence Strategic Review, for example, placed the training, equipping and advising of foreign security forces at the centre of the ongoing war against al-Qaeda. 'As US forces draw down in Afghanistan', the document detailed, 'global counter terrorism efforts will become more widely distributed and will be characterised by a mix of direct action and security force assistance' (DOD 2012, 4). Speaking to the twin security/strategic logics of Security cooperation, the Obama administration's influential Presidential

Policy Directive on Security Sector Assistance noted how security cooperation related activities were designed to accomplish more than just strengthening the security and governance capacity of partners. They also worked to 'promote partner support for US interests' including 'military access to airspace and basing rights; improved interoperability and training opportunities; and cooperation on law enforcement, counterterrorism, counternarcotics', amongst other policy areas (The White House 2013).

Despite the rollback of some Obama-era restraints on the use of force, the Trump administration has retained security cooperation as a key counterterrorism tool (Biegon and Watts 2020). The 2018 National Strategy for Counterterrorism restated the importance of 'augment[ing] the capabilities of key foreign partners to conduct critical counterterrorism activities' (The White House 2018, 23), which remained an essential component to the military response against transnational terrorist organisations. Institution-alising a process which can be traced to Obama's 'pivot to Asia', the Trump administration has recalibrated the overall strategic direction of US defence policy. According to the 2017 National Security Strategy, China and Russia 'are actively competing against the United States and our allies and partners' (The White House 2017, 25). The (re)emergence of great power competition as an organising lens for American foreign policy creates new uncertainties, including for the trajectory of remote warfare. According to Stephen Tankel (2018a), with the Trump administration 'focus[ing] more on great power conflict and rogue regimes, security cooperation with, and assistance to, allies and partners will remain critical for achieving global defense objectives.' Indeed, both the 2017 National Security Strategy and the 2018 National Defense Strategy emphasise the continued importance of such activities in tackling transnational security challenges in Africa while adding that they also have value in 'limit[ing] the malign influence of non-African powers' in the region (Department of Defence 2018, 10; see also The White House 2017, 52). Thus, whilst the immediate focus of these activities may be reoriented to reflect the new strategic focus on great power competition, security cooperation will likely remain an important instrument in the American foreign and counterterrorism policy toolbox.

Conceptualising security cooperation as a tool of remote warfare

At the core of the current debate on remote warfare is the trend towards countering security threats at a greater physical, political and strategic distance. The Oxford Research Group defines remote warfare as a 'term that describes approaches to combat that do not require the deployment of large numbers of your own ground troops' (Knowles and Watson 2018, 2). Whilst there has been somewhat of a 'pick-and-mix' approach to the way these have been catalogued, a variety of tactical practices have been studied under this label, including manned and unmanned airpower, military assistance, cyber operations, intelligence sharing, private military security contractors and special operations forces (SOF). Whilst Western states may conduct direct combat operations against shared security challenges, they do so from the air or with elite SOF units, not their conventional ground forces. The bulk of the fighting is instead delegated to local security agents whose military capacity is strengthened through security cooperation and tailored packages of operational support, often comprising embedded SOF advisors, airpower and intelligence sharing (Knowles and Watson 2018, 2-3).

When situated within this debate, security cooperation offers the attractive prospect of shaping the security situation on the ground, particularly in sites like Somalia where important, but not vital, security interests are threatened. Security cooperation can help build the capacity of partnered security agents to conduct military operations to a standard or scale that surpasses earlier capabilities, thus enabling them to better tackle shared security challenges (Biddle et al. 2017, 100). This intuitive security logic has two dimensions. On the one hand, it constitutes an effort to improve the capacity of some foreign security agents to deny transnational terrorist organisations ungoverned spaces from which to operate (Tankel 2018b, 101). On the other, it provides a means of enabling other foreign security agents to participate in coalition operations alongside or in place of American forces (Ross 2016, 96–97). What binds the security logic of security cooperation as a tool of remote warfare is that, in theory if not necessarily practice, it can 'reduce the need for US troops to do the fighting by improving the ally's ability to do this themselves' (Biddle et al. 2017, 91–92).

Beyond this, security cooperation also has strategic logics. Andrew Shapiro, former Assistant Secretary of State for Political-Military Affairs, notes how cooperation on sensitive defence issues strengthens the diplomatic relationship between the US and the recipient state, creating new patterns of cooperation, dependency and leverage (Shapiro 2012, 29–31). Whilst security cooperation does not automatically translate into influence, it can 'help tie a country's security sector to the United States' and create 'strong incentives for the recipient countries to maintain close relations, both in times of stability and in crisis' (Shapiro 2012, 30–31). Furthermore, it can help secure geographical and political-technical access, a principle which is recognised in the Joint Publication 3–20 which notes how security cooperation activities 'supports US military campaign and contingency plans with necessary access, critical infrastructure, and [partner nation] support' (Joint Chiefs of Staff 2017, v–vi). This geographical access takes multiple forms and is not restricted to just overseas basing rights. As noted in the wider literature, it can also include access to *airspace* to conduct aerial reconnaissance and strike operations; *foreign military personnel*, to build partner capacity, participate in joint counterterrorism raids and provide intelligence; and *transit*, whether this be intended to conduct military operations in a neighbouring state or to resupply US combat forces in theatre (Tankel 2018b, 105-107). In this way, the strategic logics of security cooperation can help provide the US with territorial access to partnering countries, but also a degree of technical access to those partnering security agents that, under remote warfare, do the majority of frontline fighting.

To reiterate, the provision of military assistance does not automatically translate into direct influence (Ross 2016, 94). As understood through the lens of principal-agent theory (Biddle et al. 2017), political dynamics are central to the effectiveness of partner building efforts. Questions of efficacy flow from the substantial agency loss involved in the use of these programmes, as seen in the challenges generated by adverse selection problems, interest asymmetries and the difficulties in monitoring how military training and equipment is used by recipients (Biddle et al. 2017). In contested sites of security cooperation such as Somalia, there can also be competition amongst security cooperation providers for influence, further complicating matters. As one interviewee involved with British partner building efforts in Somalia put it, 'when you're there as a team of 15 you don't have automatic influence [...] so you need time to build relationships instead. You're there competing with other internationals for influence' (quoted in Watson and Knowles 2019, 3). Even in such situations, however, security cooperation activities can help generate the different forms of access outlined above. As Knowles and Watson document, for a comparatively modest investment in manpower and resources the United Kingdom was able to secure access into the operations and intelligence room at AMISOM via its partner building efforts in Somalia: 'a high level of access - which could lead to more effective partnerships in the future' (Knowles and Watson 2018, 4). In addition to the political dynamics intrinsic *within* the delivery of security cooperation which impact the effectiveness of associated programmes, the political context informing the 'principal's' decision to provide the 'agent' with assistance are thus also worthy of consideration.

US security cooperation as remote warfare in the Horn of Africa

The external training, equipping and advising of African security forces is not new. European powers relied heavily on locally-raised militaries to augment their own ground forces throughout the age of empire (Johnson 2017, 173–194). During the Cold War, the US government provided military assistance to states across the Horn of Africa (Kuzmarov 2017). The region was a site of acute East-West competition, with both superpowers active in advancing their respective ideological and geopolitical interests across the region (Makinda 1982, 98–101). The provision of military assistance had both security and strategic logics. It was intended to help maintain access to air and naval facilities in Ethiopia and later Somalia; defend the internal stability of partner governments; and maintain the openness of the strategically important Bab-el-Mandeb waterway, a key artery of global trade (Lewis 1987, 3). This effort to manage security challenges in the Horn of Africa from 'over the horizon' was given further impetus by the deaths of eighteen Army Rangers during the 1993 Battle of Mogadishu, popularly known as the 'Black Hawk Down' incident. As Robert Patman (2015) has argued, the resultant 'Somalia Syndrome' generated a profound scepticism about intervening on the ground in humanitarian crises, shaping later remote warfare campaigns in Africa.

Following the 9/11 attacks, Bush administration officials feared that al-Qaeda's senior leadership would relocate to the Horn of Africa following their expulsion from Afghanistan (Ryan 2019, 82–83). Based at Camp Lemonnier in Djibouti, the Combined Joint Task Force-Horn of Africa (CJTF-HOA) was created in October 2002 to coordinate counterterrorism activities throughout the region with a strong focus on building partner capacity and civil-military operations (Ryan 2019, 85-88). Following its breakaway from the Islamic Courts Union in 2006 against the backdrop of the US-backed Ethiopian invasion of Somalia, al-Shabaab emerged as the principal target of CJTF-HOA's activities. This al-Qaeda affiliated group has fought an effective insurgency against the Federal Government of Somalia and the African Union Mission in Somalia (AMISOM), the latter which was created in 2007 to support the nominal Somali state. Al-Shabaab has at times controlled large swathes of territory in central and southern Somalia, conducted terrorist attacks in neighbouring Kenya and Ethiopia, and infiltrated Somalia's security and intelligence services (Reno 2018, 502–503).

Beginning in George W. Bush's presidency, against the advice of local partners to keep a 'low profile' in order to minimise the risk to peacekeeping contingents (Wikileaks 2007a), successive US administrations have utilised Security cooperation alongside other remote practices of intervention.[5] In 2007 the DOD's Joint Special Operations Command (JSOC) was authorised to conduct air strikes from manned/unmanned aircraft and conduct 'kill/capture' SOF raids against al-Shabaab's senior leadership. By January 2017, between 32–36 covert strikes are reported to have been conducted, with the first drone strike reportedly occurring in June 2011 (the Bureau of

---

[5]   American SOF have also been active in Somalia from 2007 onward providing local security agents training, advice, mission planning, communication support and medical expertise (Stewart 2014). They have also conducted covert kill-capture raids against Al-Shabaab's leadership (Mazzetti, Gettleman and Schmitt 2016).

Investigative Journalism 2017). Whilst disruptive, such strikes formed a small part of a larger package of intervention: 'American strategy for containing and ultimately defeating al Shabaab relie[d] on AMISOM and the Somali National Army' (Zimmerman, Meyer, Lahiff and Indermuehle 2017). This illustrates the centrality of security cooperation to this particular remote warfare campaign.

According to data provided by the Security Assistance Monitor (2019), Somalia was allocated at least $248.6 million in military assistance in the period between FY 2006–2018.[6] In a 2009 diplomatic cable sent from the American embassy in Ethiopia, concerns were expressed about providing military support to the fledging Somali Transitional Federal Government without strengthening its capacity to govern and provide public services because such actions 'raises US involvement in the morass of a Somali civil war in the name of counterterrorism' (Wikileaks 2009a). In this spirit, it was not until 2013 that the Obama administration lifted restrictions on the provision of defence equipment and services to the Somali army (Ross 2018), with the effort to build capacity in the Somali National Army (SNA) gaining further momentum following the April 2015 announcement of the Guulwade (Victory) Plan which aimed to create a 10,900 strong person security force capable of facilitating AMISOM's withdrawal from Somalia (Reno 2018, 500). Despite these efforts, the SNA remained chronically undermanned, poorly led and badly equipped (Matisek 2018, 278–279). It was, in Paul Williams' assessment, 'an army in name only, largely confined to defensive and localised operations, unable to undertake a coherent national campaign, and often reliant on [others] for protection, securing its main supply routes, logistics support and casualty evacuation' (Williams 2019, 2). The Lightning 'Danab' advanced infantry company, one of the few comparative successes of US partner building activities for example, generally operated separately from the SNA (Williams 2019, 2), and was reportedly insulated from the influence of some Somali government officials (Reno 2018, 508–509). Reflecting these and a myriad of other political, contextual and operational challenges (Williams 2019), the focus of American security cooperation efforts in the Horn of Africa concentrated on AMISOM.

The six AMISOM contributing states listed in Figure 1 received $1.28 billion in military assistance between FY 2006-2018 (Security Assistance Monitor 2019).

---

[6]   This figure has been calculated by subtracting peacekeeping operations funding from the total military assistance allocated to Somalia during this period. As the Security Assistance Monitor notes, whilst 'the US has historically appropriated Peacekeeping Operations assistance to Somalia with the intent to support both the Somali National Forces and AMISOM [...] [the] US Government reports do not provide details about how [Peacekeeping Operations] amounts are divided between the two security providers' (Chwalisz 2014).

Figure 1: AMISOM troop contributing states.[7]

State	Year joined AMISOM	Peak AMISOM troop contribution
Burundi	2007	5,400
Djibouti	2011	1,800
Ethiopia	2014	4,400
Kenya	2012	4,300
Sierra Leone	2013	850
Uganda	2007	6,200

This assistance was provided both directly to AMISOM contributing states and indirectly via the United Nations Support Office in Somalia (Ross 2018). Examples of the first form of assistance include the use of the counterterrorism oriented Section 1206/Section 3333 programme ($730.5 million) and the Counterterrorism Partnership Fund ($59 million).[8] As a region, East Africa was also allocated $275.9 million in Counterterrorism Partnership Fund assistance between FY 2015-2016 and $112.2 million in Section 1207(n) Transitional Authority funds between FY's 2012-2014 (Security Assistance Monitor 2019).[9] Beyond this, AMISOM was also allocated at least $2 billion in funding via the State Department's Peacekeeping Operations account (Security Assistance Monitor 2019). According to a 2014 White House factsheet, $512 million had also been committed to support AMISOM

---

[7]   This table is modified from (Williams 2018, 174).

[8]   Prior to its consolidation into the larger Section 333 authority as part of the 2017 National Defense Authorization Act (NDAA), the 'Global Train and Equip Authority' was used to build the capacity of foreign military, maritime and border forces to conduct counterterrorism operations and support US coalition missions. For a more detailed discussion of this authority's history and purpose, see (Ryan 2019, 153–156). Authorised in the FY2015 NDAA, the Counterterrorism Partnership Fund was intended to build partner capacity principally in frontline states in Africa and the Middle East, with a focus on intelligence, surveillance and reconnaissance, border security, airlift, counter-improvised explosive device capabilities and peacekeeping (Office of the Under Secretary of Defense 2016, 2).

[9]   The Section 1207(n) Transitional Authority was a three-year transnational authority, attached to the Global Security Contingency Fund in the FY2012 NDAA, which supported counterterrorism operations in the Arabian Peninsula and the Horn of Africa. It had two specific goals: 'enhance the capacity of the national military forces, security agencies serving a similar defence function, and border security forces of Djibouti, Ethiopia, and Kenya to conduct counterterrorism operations against al-Qaeda, al-Qaeda affiliates, and al Shabaab' on the one hand, and '[t]o enhance the ability of the Yemen Ministry of Interior Counter Terrorism Forces to conduct counterterrorism operations against al-Qaeda in the Arabian Peninsula and its affiliates' on the other (Serafino 2014, 5 FN).

via 'pre-deployment training, provision of military equipment, and advisors on the ground' (The White House 2014).

Figure 2: US military assistance to AMISOM contributing states FY2006-2018 in millions of $.[10]

State	Total US Military Assistance	Section 1206/ Section 3333	Counterterrorism Partnership Fund
Burundi	$53.2	$34.7	-
Djibouti	$77.4	$37.8	-
Ethiopia	$121.5	$67.4	$18.7
Kenya	$628.3	$354.4	$31.4
Sierra Leone	$27.9	$0.1	-
Uganda	$373.8	$236.1	$8.9
Total	$1,282.1	$730.5	$59

Consistent with the use of security cooperation to enable partners to participate in coalition operations, these funds were allocated to plug key gaps in their recipients' counterterrorism capacity. Section 1207(n) funds, for example, were intended to build the capacity of 'Djibouti, Ethiopia, and Kenya to conduct counterterrorism operations against al-Qaeda, al-Qaeda affiliates, and al Shabaab' (Serafino 2014, 5 FN). Likewise, CTFP funds were requested to improve AMISOM contributors' intelligence, surveillance and reconnaissance, counterterrorism interdiction, counter-improvised explosive device and command and control capabilities (Office of the Under Secretary of Defense 2016, 5–6). Such narrow focus on 'plugging in' greater tactical competences without building the institutional and logistical architectures to support them has raised questions about the sustainability of these gains once the funding taps are turned off (Ross 2018).

As noted in the wider literature, the efficacy of security cooperation activities is contentious (Biddle et al. 2017; Matisek 2018; Reno 2018). In Somalia, there has been an overemphasis on building the tactical capability of local security forces at the expense of the political and institutional reforms required for long-term conflict resolution (Williams 2019, 13), as well as an inattention to wider strategy (Ross 2018). Compounding these failures of

---

[10]   This table has been generated from data from (Security Assistance Monitor 2019). The total US military assistance figure includes support provided through both Pentagon-managed security cooperation programmes and State Department-managed Security Assistance programmes.

execution are the structural limits of what is possible for security cooperation to accomplish in conditions of state collapse. These result from the misalignment of interests between the US and various local actors (Reno 2018, 505; Williams 2019, 15–17; Matiesk 2018, 278–279). Local partners retain their own agency, and in the case of Somalia, have lacked the political will or incentive to realign their behaviours according to America's security preferences. According to an unnamed Pentagon official, 'eliminating al-Shabaab is the easy part; the hard part is getting the institutions of Somalia to work' (quoted in Matisek 2018, 278). These barriers are consistent with the principal-agent issues that characterise the use of this particular policy tool. The very distance between the donor-as-principal and the recipient-as-agent that enables security cooperation to serve as a means of remote warfare also undermines its efficacy as a security tool (Biddle et al. 2017).

Notwithstanding these barriers to the conversion of military support into desired political outcomes, an expanded focus on the strategic logics of security cooperation opens up an alternative calculus to qualify the well-documented failures of these activities. Consistent with our earlier conceptualisation of the security logics of security cooperation, despite agency losses and aid misappropriation, security cooperation has enabled American policymakers to exert at least some influence on the ground in the region whilst continuing to distance conventional US ground forces from the bulk of frontline fighting. The training and equipment provided through the Section 1206 authority improved the capacity of frontline states such as Ethiopia and Kenya to better police their border and coastal regions prior to joining AMISOM, helping limit al-Shabaab's freedom of movement. Security cooperation also incentivised and facilitated AMISOM troop contributions to fight in Somalia itself. Contributing troops to AMISOM enabled the Uganda Peoples' Defence Force to access both US peacekeeping- and counter-terrorism-orientated funding, training and assistance (Williams 2018, 176). Similarly, as a perquisite for its participation, the Government of Burundi 'compiled a 20-page list of requests that it considered necessary to join AMISOM, including trucks and bulldozers, aircraft, and helicopters as well as office supplies, sleeping bags, personal equipment, and optical equipment such as night vision goggles' (Williams 2018, 177). Whilst other political, institutional and normative considerations influenced the decision of the six AMISOM contributing states to provide troops to fight in Somalia, increased receipt of US military assistance alongside other avenues of financial support was often, but not always, a motivating factor (Williams 2018).

Additionally, discussions in leaked embassy cables and public press releases illustrate the ways in which security cooperation initiatives thicken political partnerships with key regional states. In 2007, the US ambassador to Kenya discussed 'synchronising efforts' across the Horn of Africa, through a 'multipronged approach involving continued military and security actions' with other diplomatic and development efforts. He also 'stressed the need for American officials and contractors to visit Somalia', as 'such visits were essential both for operations and to effectively publicise both within Somalia and the region the good work' the US was carrying out (WikiLeaks 2007b). In Ethiopia, the provision of aircraft maintenance was argued to be 'critical to continuing a viable (military-military) relationship with a proven partner in the war on terrorism' (WikiLeaks 2007c). Diplomatic staff based in Addis Ababa expressed concern that, due to the repeated failures to repair two Ethiopian operated C-130s military transport aircraft and the anticipated closure of the US-funded Ethiopian Defense Command and Staff College, some within the Ethiopian military were aiming 'to make China, and to a lesser extent Israel, their major military relationship' (Wikileaks 2007d). Security cooperation activities also strengthened cooperation between regional partners, including on sensitive areas such as intelligence (Hurd 2019), and provided the US with technical access to partnering security agents.

In 2016, following the completion of the first annual military-to-military engagement event African Partnership Flight, a US Air Force spokesperson explained that bringing together participants from the Kenyan and Ugandan air forces under US instruction would 'build enduring relationships with (US) partner countries.' Speaking to the collaborative spill-over effects of security cooperation, the spokesperson further noted that through such activities the US had '[built] a partnership and friendship that has helped open the door for further engagement, knowledge sharing and interoperability between our forces' (quoted in Chavez 2016). A similar logic punctuates the US Army's annual Justified Accord Exercise, initiated in 2017, which functions to improve the capacity of regional forces to support AMISOM and develop intra-personal relationships with, and access, to local forces. As Lapthe C. Flora, the then US Army Africa deputy, put it in 2019:

> 'I cannot overemphasise the importance of exercises like Justified Accord [...] They not only contribute to the readiness of African nations and peacekeeping operations, but they also provide valuable opportunities to work together and create professional relationships and friendships' (quoted in Valley 2019).

Finally, whilst it is difficult to document an exact 'transmission belt' between an increase in security cooperation and the production of access, the increase in security cooperation activities to combat al-Shabaab has paralleled the rollout of military installations across the Horn of Africa. Officially, the US operates only one military base in Africa, Camp Lemonnier in Djibouti

(Moore and Walker 2016, 686). Around this, however, a constellation of smaller 'cooperative security locations' orientated around drone, SOF and contractor assemblages have been operated, with suspected locations in Ethiopia, Kenya, the Seychelles, Somalia, South Sudan and Uganda (Moore and Walker 2016; Turse 2018). In the case of the Seychelles, there is evidence to suggest that military assistance was used to thicken the US' bilateral partnership with the host government following the basing of a small fleet of unarmed MQ-9 Reapers on the island to conduct anti-piracy and surveillance missions.

During an August 2009 meeting with AFRICOM commander General William Ward, Seychelles President Michel noted that his island was an 'aircraft carrier in the middle of the Indian Ocean without the planes' and welcomed 'this resurgence of American military activity in the Seychelles' (Wikileaks 2009b). Following the initial use of these facilities in September 2009, the overall level of US military assistance rose from $251,299 in FY 2010 (an accounting period which began on 1 October 2009) to $893,244 in FY2011 (Security Assistance Monitor 2019). Consistent with General Ward's expressed commitment to strengthen bilateral military relations and improve the capacity of the islands' coastguard (Wikileaks 2009b), $535,000 was allocated in this year via the State Department Foreign Military Financing programme for Metal Shark patrol boats and 'Secure Video and Data Link equipment' (Department of State 2014, 11). Following the suspected suspension of drone operations from this base at some point in 2012 (Moore and Walker 2016, 696), overall military assistance to the Seychelles declined from $627,580 in FY2012 to $464,555 in FY2013 and $208,224 in FY2014 (Security Assistance Monitor 2019).

Conclusion

Security cooperation offered the Bush, Obama and Trump administrations an attractive means of 'squaring the circle' on the use of military force. As a tool of remote warfare, it allows planners to exert limited influence 'on the ground' in complex overseas security environments, but without deploying large numbers of their own 'boots on the ground' to conduct frontline fighting. The security logic that is foregrounded in much of the study of these activities is an intuitive component of this feature of US military interventionism. However, as we have argued, this sits alongside a parallel set of strategic logics. Security cooperation has helped secure various forms of geographic and technical-political access, including on matters of basing, airspace and transit rights; thickened political partnership; and helped create patterns of cooperation, influence and leverage.

In consideration of emerging debates on the effectiveness of remote warfare, we have highlighted the need to account for the dual security and strategic logics of policy tools like security cooperation. The intersecting features of remote warfare, as expressed through its kinetic and non-kinetic dimensions, are illuminated in the recent history of US policy in the Horn of Africa. In Somalia, the US has consistently used security cooperation alongside other remote practices of intervention. The ability of the US to confront al-Shabaab directly or indirectly has been contingent on Washington's capacity to secure access and partnerships in the region. The significance of security cooperation in a country like Somalia needs to be understood against the backdrop of the conditions that elicited the turn toward remote warfare on the part of the US and other agents. Absent alternatives, security cooperation programmes have provided a pathway to continued intervention, the 'remoteness' of which applies only to the intervening actor, not the local communities for whom political violence is intimate. This is not to claim that US intervention in the Horn of Africa has been successful or that its failings are fixable using more or different configurations of remote warfare practices. Rather, it is to suggest that the dynamics of remote warfare need to be analysed holistically, and in conjunction with the twin security and strategic purposes they serve.

References

Biddle, Stephen, Julia Macdonald, and Ryan Baker. 2017. 'Small footprint, small payoff: The military effectiveness of security force assistance.' *Journal of Strategic Studies*, 41(2): 89–142.

Biegon, Rubrick, and Tom Watts. 'When ends Trump means: continuity versus change in US counterterrorism policy.' *Global Affairs*, 6(1): 37–53.

Chavez, Evelyn. 2016. 'African Partnership Flight Kenya comes to a close.' *Royal Air Force Mildenhall*. 30 June. Accessed 26 November 2019. https://www.mildenhall.af.mil/News/Article-Display/Article/820775/african-partnership-flight-kenya-comes-to-a-close/

Chwalisz, Natalie. 2014. 'United States Peacekeeping Operations (PKO) assistance in Somalia.' *Security Assistance Monitor*. 4 April. Accessed 10 October 2017. http://www.securityassistance.org/blog/united-states-peacekeeping-operations-pko-assistance-somalia

Department of State. 2014. *Report on the Use of Foreign Military Financing, International Military Education and Training, and Peacekeeping Funds.* Accessed 15 November 2019. http://securityassistance.org/sites/default/files/Section 7010 Reports as of 6-30-15.pdf

DOD. 2006. *Quadrennial Defense Review Report 2006*. Accessed 16 September 2017. http://archive.defense.gov/pubs/pdfs/QDR20060203.pdf

DOD. 2012. *Sustaining Global Leadership: Priorities For 21st Century Defence*. Accessed 3 March 2017. http://archive.defense.gov/news/Defense_Strategic_Guidance.pdf

Gates, Robert. 2007. Secretary of Defense Speech. US DOD. 26 November. Accessed 16 September 2017. http://archive.defense.gov/speeches/speech.aspx?speechid=1199

House of Representatives. 1963. Foreign operations appropriations for 1964: hearings before a subcommittee of the Committee on Appropriations, House of Representatives, eighty-eighth Congress, first session. Accessed 15 November 2019. file://catalog.hathitrust.org/Record/006212313

Hurd, Christopher. 2019. 'East Africa DMI Conference: Information Sharing Through Lasting Relationships.' AFRICOM. 22 April. Accessed 26 November 2019. https://www.africom.mil/media-room/article/31782/east-africa-dmi-conference-information-sharing-through-lasting-relationships

Johnson, Rob. 2017. *True to Their Salt: Indigenous Personnel in Western Armed Forces*. Oxford: Oxford University Press.

Joint Chiefs of Staff. 2017. *Joint Publication 3-20: Security cooperation*. Accessed 10 October 2017. http://www.dtic.mil/doctrine/new_pubs/jp3_20_20172305.pdf

Knowles, E., and Watson, A. 2018. *Remote Warfare: Lessons Learned from Contemporary Theatres*. London: Oxford Research Group. Accessed 11 December 2018. https://www.oxfordresearchgroup.org.uk/Handlers/Download.ashx?IDMF=4026f48e-df60-458b-a318-2a3eb4522ae7

Kolko, Gabriel. 1988. *Confronting the third world: United States foreign policy, 1945–1980*. New York: Pantheon.

Kuzmarov, Jeremy. 2017. 'American Military Assistance Programs Since 1945.' *Oxford Research Encyclopaedia of American History*. https://oxfordre.com/americanhistory/abstract/10.1093/acrefore/9780199329175.001.0001/acrefore-9780199329175-e-346

Lewis, William. 1987. 'US Military Assistance to Africa.' *CSIS Africa Notes* 75 (1987): 1-7. Accessed 18 May 2019. https://csis-prod.s3.amazonaws.com/ s3fs-public/legacy_files/files/publication/anotes_0887.pdf

Makinda, Samuel. M. 1982. 'Conflict and the Superpowers in the Horn of Africa.'*Third World Quarterly*, 4(1): 93–103.

Matisek, Jahara. 2018. 'The crisis of American military assistance: strategic dithering and Fabergé Egg armies.' *Defense and Security Analysis*, 34(3), 267–290.

Mazzetti, Mark, Jeffrey Gettleman, and Eric Schmitt. 2016. 'In Somalia, US Escalates a Shadow War.' *New York Times*. 16 November. Accessed 16 September 2017. https://www.nytimes.com/2016/10/16/world/africa/obama-somalia-secret-war.html

Moore, Adam, and James Walker. 'Tracing the US military's presence in Africa.' *Geopolitics*, 21(3): 686–716.

Oberdorfer, Don. 1977. 'US Offers Military Aid to Somalia.' *Washington Post*. 26 July. Accessed 24 November 2019. https://www.washingtonpost.com/ archive/politics/1977/07/26/us-offers-military-aid-to-somalia/43249aa8-57e5-45d3-a55c-c4eec934da9a/

Office of the Under Secretary of Defense. 2016. *Counterterrorism Partnerships Fund: Department of Defense Budget Fiscal Year (FY) 2017*. Accessed 16 September 2017. http://comptroller.defense.gov/Portals/45/ Documents/defbudget/fy2017/FY2017_CTPF_J-Book.pdf

Patman, Robert. 2015. 'The roots of strategic failure: The Somalia Syndrome and Al-qaeda 's path to 9/11.' *International Politics*, 52(1): 89–109.

Reno, William. 2018. The politics of security assistance in the horn of Africa. *Defence Studies*, 18(4): 498–513.

Ross, Jay. 1981. 'Kissinger: Confront Soviets in E. Africa Kissinger Advocates More Active Anti-Soviet Role in East Africa.' *Washington Post*. 4 January. Accessed 14 November 2019. https://www.washingtonpost.com/archive/ politics/1981/01/04/kissinger-confront-soviets-in-e-africa-kissinger-advocates-more-active-anti-soviet-role-in-east-africa/e6974d2b-82ae-4ab1-ac2a-698c84e14776/

Ross, Tommy. 2016. 'Leveraging Security cooperation as Military Strategy.'*The Washington Quarterly*, 39(3): 91–103.

——. 2018. The Dangers of Incoherent Strategy: Security Assistance in Somalia, 2009–2018. *Texas National Security Review*. 20 November. Accessed 25 June 2019. https://tnsr.org/roundtable/policy-roundtable-the-pros-and-cons-of-security-assistance/

Ryan, Maria. 2019. *Full Spectrum Dominance: Irregular Warfare and the War on Terror*. Stanford: Stanford University Press.

——. 2020. '"Enormous Opportunities" and "Hot Frontiers": Sub-Saharan Africa in US Grand Strategy, 2001-Present.' *The International History Review*, 42(1): 155–75.

Security Assistance Monitor. 2019. Data: Programs FY 06-18. Accessed 25 June 2019. http://securityassistance.org/data/program/military/country/2006/2018/all/Global/

Serafino, Nina. 2014. Global Security Contingency Fund: Summary and Issue Overview. Library of Congress Congressional Washington DC Research Service.
Accessed 3 March 2017. https://fas.org/sgp/crs/row/R42641.pdf

Shapiro, Andrew. J. 2012. 'New Era for US Security Assistance.' *The Washington Quarterly*, 35(4): 23–35.

Stewart, Phil. 2014. 'US discloses secret Somalia military presence, up to 120 troops.' *Reuters.* 3 July.   https://www.reuters.com/article/us-usa-somalia/exclusive-u-s-discloses-secret-somalia-military-presence-up-to-120-troops-idUSKBN0F72A820140703

Stokes, Doug, and Kit Waterman. 2017. 'Beyond balancing? Intrastate conflict and US grand strategy.' *Journal of Strategic Studies,* 41(6): 1–26.

Tankel, Stephen. 2018a. 'Introduction: The Future of Security Assistance.'*Texas National Secuity Review.* 20 November. Accessed 25 June 2019. https://tnsr.org/roundtable/policy-roundtable-the-pros-and-cons-of-security-assistance/

Tankel, Stephen. 2018b. *With us and Against us: how America's partners help and hinder the war on terror*. New York: Columbia University Press.

The Bureau of Investigative Journalism. 2017. 'Somalia: Reported US covert actions 2001-2016.' Accessed 10 October 2019. https://www.thebureauinvestigates.com/drone-war/data/somalia-reported-us-covert-actions-2001-2017

The White House. 2013. 'Fact Sheet: US Security Sector Assistance Policy.' Accessed 11 August 2018. from https://obamawhitehouse.archives.gov/the-press-office/2013/04/05/fact-sheet-us-security-sector-assistance-policy

———. 2014. 'Fact Sheet: US Support for Peacekeeping in Africa.' Accessed 11 August 2018. https://obamawhitehouse.archives.gov/the-press-office/2014/08/06/fact-sheet-us-support-peacekeeping-africa

———. 2017. *National Security Strategy of the United States of America, 2017*. Accessed 11 August 2018. https://www.whitehouse.gov/wp-content/uploads/2017/12/NSS-Final-12-18-2017-0905.pdf

———. 2018. *National Strategy for Counterterrorism of the United States of America*. Accessed 12 January 2019. https://www.whitehouse.gov/wp-content/uploads/2018/10/NSCT.pdf

Turse, Nick. 2018. 'US Military Says It Has A "Light Footprint" In Africa. These Documents Show A Vast Network Of Bases.' *The Intercept*. 1 Decmber. Accessed 26 November 2019. https://theintercept.com/2018/12/01/u-s-military-says-it-has-a-light-footprint-in-africa-these-documents-show-a-vast-network-of-bases/

Valley, Ian. 2019. *Exercise Justified Accord 2019 concludes*. United States Army Africa. https://www.usaraf.army.mil/media-room/article/29275/exercise-justified-accord-2019-concludes#:~:text=

Watson, Abigail, and Emily Knowles. 2019. 'Improving the UK offer in Africa: Lessons from military partnerships on the continent.' London: Oxford Research Group. Accessed 5 November 2019. https://www.oxfordresearchgroup.org.uk/Handlers/Download.ashx?IDMF=94a8e650-09a0-408e-bc0d-3c0610c10d5e

Watts, Tom, and Rubrick Biegon. 2017. *Defining Remote Warfare: Security cooperation*. Accessed 11 May 2019. http://www.oxfordresearchgroup.org.uk/sites/default/files/RCP-Security_Cooperation.pdf

————. 2019. Conceptualising Remote Warfare: The Past, Present, and Future. Accessed 26 May 2019. https://www.oxfordresearchgroup.org.uk/conceptualising-remote-warfare-the-past-present-and-future

White, Taylor. 2014. 'Security cooperation: How It All Fits.' *Joint Forces Quarterly*, 72: 106–108.

Williams, Paul D., 2018. 'Joining AMISOM: why six African states contributed troops to the African Union Mission in Somalia.' *Journal of Eastern African Studies*, 12(1): 172–192.

————. 2019. 'Building the Somali National Army: Anatomy of a failure, 2008–2018.' *Journal of Strategic Studies,* 43(3): 366–391.

Wikileaks. 2007a. 'Ethiopian FM Hails Cooperation, But Seeks Lower US Military Profile, In Somalia.'ID:07ADDISABABA89_a. 12 January. Accessed 26 November 2019. https://wikileaks.org/plusd/cables/07ADDISABABA89_a.html

————. 2007b. 'East Africa Regional Strategic Initiative (RSI), March 16-17.' ID:07DJIBOUTI425_a. 10 April. Accessed 26 November 2019. https://wikileaks.org/plusd/cables/07DJIBOUTI425_a.html

————. 2007c. 'Section 1206 Proposal Final Coordination.' ID:07ADDISABABA852_a. 19 March. Accessed 26 November 2019. https://wikileaks.org/plusd/cables/07ADDISABABA852_a.html

————. 2007d. 'Ethiopia: Broken Promises, Unmet Expectations -- Fixing Mil-To-Mil Relations.' ID: 07ADDISABABA1535_a. 21 May. Accessed 26 November 2019. from https://wikileaks.org/plusd/cables/07ADDISABABA1535_a.html

————.2009a. 'Ethiopia, Somalia, AMISOM, and the Way Forward.' ID: 09ADDISABABA1186_a. 19 May. Accessed 26 November 2019. https://wikileaks.org/plusd/cables/09ADDISABABA1186_a.html

—————.2009b. 'Seychelles: General Ward/US Africa Command Visit 19 August Helps Cement New, Closer Relationship.' ID: 09PORTLOUIS271_a. 8 September. Accessed 26 November 2019. https://wikileaks.org/plusd/cables/09PORTLOUIS271_a.html

Zimmerman, Katherine, Jacqulyn Meyer Kantack, Colin Lahiff, and Jordan Indermuehle. 2017. 'US Counterterrorism Objectives in Somalia: Is Mission Failure Likely?' American Enterprise Institute. Accessed 16 September 2017. https://www.criticalthreats.org/analysis/us-counterterrorism-objectives-in-somalia-is-mission-failure-likely

10

The Limitations and Consequences of Remote Warfare in Syria

SINAN HATAHET

Remote warfare aims to reduce the risks and costs of traditional military intervention and externalise the burdens of warfare to local partners and non-state actors. However, in reality, this practice comes with high risks. Non-state actors often pursue their own agenda and sometimes act against the wishes and advice of their backers. Delegating the strategic, operational and tactical burden of a foreign policy to local partners often comes at the expense of control and could easily escalate to new levels of violence. In the past nine years, Syria has been transformed into a theatre of complex remote warfare, waged by regional actors against neighbouring rivals, international powers against both rogue states and armed groups, and transnational terrorist organisations against incumbent states and local populations. These conflictual and unreconcilable foreign agendas have not only fuelled the ongoing war between Assad regime and the rebels, but they have also further destabilised the region, creating more animosity and mistrust among the different involved actors.

The chapter begins by providing a background to the conflict. The chapter then recounts the primary states engaged in remote warfare in Syria, their objectives and models of intervention. After this, it delves into the different types of interactions these states had with their respective Syrian partners or proxies. Following this overview, the chapter investigates how Syrian armed groups exercised their agencies, established their governance structures, and how these choices impacted the support they received from their backers and vice-versa. Finally, the chapter concludes with a sketch of the possible outcomes of the Syrian conflict on the armed groups' roles in Syria and beyond with the eventual withdrawal of their backers.

Background to the conflict in Syria

Following Assad's violent crackdown on civilians in 2011, some defected military officers and militants staged an armed insurgency against the Syrian Army, which is still ongoing. Encouraged by their constituencies and the flow of financial and material foreign assistance, these armed groups developed governance structures and claimed political and security roles (Bojicic-Dzelilovic 2018). Meanwhile, the Syrian state security and military apparatus suffered from significant human and material losses and had to give up control over large swaths of land to ensure the regime's survival in the capital and coastal regions. Despite the opposition's best efforts to fill the void left by the state withdrawal from northern and eastern Syria, transnational extremist actors such as al-Qaeda and the Islamic State in Iraq and the Levant emerged and seized the opportunity to compete with grassroots movements (Rich and Conduit 2015).

In reaction to the growing instability in Syria, some regional and global state actors were concerned with the threats emanating from the country and decided to partner with local armed groups to collectively face these threats with a minimum military engagement on their behalf. However, what started as a security concern for some powerful states evolved to an objective to play a leading role in shaping the future of Syria. Another host of countries found an opportunity in the civil conflict either to re-enforce their influence over Damascus or to challenge Assad's authority and eventually induce a regime change. This group, too, chose to support non-state actors.

Directly impacted by the ongoing events of the Syrian conflict, regional powers such as Iran and Turkey not only backed local actors, but also deployed their own soldiers to the battlefield. Both were impelled to increase their footprint in the war when the geopolitical implications of their absence grew too costly to sustain. In both cases, their direct intervention was deemed unavoidable, justifiable and legitimate in the eyes of their domestic public opinion or their political elite. This same logic did not apply in the case of international powers, such as the US, France and Britain. There are important reasons for this. First, Syria is too distant to create a sense of urgency or emanating threat on the domestic level. Second, the West's military interventions in Afghanistan and Iraq resulted in war fatigue; there was simply no domestic appetite or support for waging new wars. Consequently, withdrawing forces from the Middle East became an essential part of their political discourse, finding justifications for further engagement even for moral imperatives seemed too challenging to achieve.

Eventually, the emergence of the Islamic State in Iraq and the Levant and the increase of terrorist attacks perpetrated by the group across Europe and the US established motives for the West's involvement in Syria (Kagan 2016). However, even then, any direct military intervention in Syria was still perceived domestically as financially, materially and politically too costly. Alternatively, the US along with 81 other countries founded the Global Coalition Against Daesh in 2014. To defeat the terrorist organisation, the Coalition adopted a dual strategy of counterterrorist military air strikes, and advising, training, and equipping local partners to plan and execute ground operations against Islamic State in Syria and Iraq. The US response to the threat posed by Islamic State demonstrates the will to achieve two objectives: first, burden-sharing with other states; second, maintaining a light footprint in the region through maximum use of technologically advanced weapons and minimum deployment of boots on the ground.

Russia, on the other hand, was not subject to the same imperatives or logic. The collective trauma of the Soviet defeat in Afghanistan was a distant memory, and, under Putin's leadership, Moscow was eager to reclaim a more assertive role on the international scene. Nonetheless, Russia, too, used the emergence of Islamic State to justify its intervention in Syria, and similarly it imprinted its military footprint through remote warfare, supporting Assad loyalist forces and mostly engaging its air forces only. The Russian approach was pragmatic and mostly motivated by financial constraints.

At the peak of the Syrian conflict in 2015, a myriad of regional and international powers were actively or indirectly engaged on the battlefield. Their different agenda and objectives further escalated tensions between them, disrupting traditional alliances and creating room for the establishment of new opportunistic arrangements between unlikely partners. An expected outcome of these dynamics was the emergence of different alliances and hence, competing projects, paving the way for remote warfare led by foreign powers, and fought by local forces to overcome their adversaries. Meanwhile, Syrian armed groups benefited from these growing hostilities to expand their agencies and to forcibly claim authority over larger populations with a devastating impact on civilian lives.

Foreign actors

The influence of foreign intervention in the Syrian conflict is neither unique nor peculiar. With 71 percent of civil wars recording at least one intervention, foreign intervention in civil wars is the rule rather than the exception (Achen and Snidel 1989). In the Syria case, foreign intervention has involved the transfer of money, arms and foreign militants (Hinnebusch 2017). From the outset of the conflict, both loyalist and opposition forces demonstrated a strong desire to rely on foreign assistance to overcome their adversaries. The Syrian opposition mainly sought financial support and weapons to wage their military operations against the Syrian army and has welcomed foreign volunteers, especially at the beginning of the conflict (Krieg 2016). Similarly, Assad pursued military and intelligence assistance to compensate for his losses on the ground, and eventually allowed the formation of foreign battalions to fight alongside his forces (Fulton 2013). What is unique about the Syrian case is not the phenomena of foreign intervention per se, but rather the extent to which international backers controlled the course and actions of local clients in waging remote warfare with minimum human and financial cost on their part.

The ability of foreign powers to play such a role could be partially explained by geopolitics and the strategic value of Syria in a polarised region festering with rivalries. However, such an impact would not have been possible if it was not for the complex Syrian demography and the social rifts between religious and 'secular,' Sunnis and Alawites, Arabs and Kurds (Phillips 2015). These divisions were often craftily manipulated by emerging and traditional regional powers and between the West and Russia.

The conflict in Syria started with two main camps: one in support of the opposition consisted of the US, EU, France, UK, Turkey, Saudi Arabia and Qatar, and the other in support of the regime, mainly Russia, Iran and to a lesser degree China. For the US and Europe, their initial pro-opposition stance is derived from a broader support policy towards the Arab Spring revolutions. This position changed later as Islamic State emerged in eastern Syria and western Iraq and they adopted a counterterrorism policy to deal with the Jihadist threat in the region and elsewhere (Krieg 2016). In contrast, Russia's initial diplomatic support to Assad was motivated by the need to preserve its international status after NATO's perceived 'betrayal' in Libya (Katz 2011). Previously engaged in preventing the fall of Assad, Putin escalated his investment in the Syrian regime only in 2015 when he identified the conflict as an opportunity to upgrade his role in the Middle East. As for the Arab oil-rich monarchies led by Saudi Arabia, their intervention in the conflict aimed to prevent Syria from being open to the Iranian power projection in the Mediterranean region. Inversely, Tehran viewed the Syrian uprising as a direct threat to its regional presence and identified the opposition as the tools of regional rivals and as agents of the US and other Western powers (Rabinovich 2017). For Turkey and Qatar, the Syrian uprising represented an opportunity to create a new political structure in the Middle East, a post-Arab Spring populist order led by both as supporters of revolutionary forces (Pala 2015).

However, as the military operations intensified, new local and regional dynamics emerged, creating new alliances and alignments among these foreign actors and changing the nature of their interventions as well as their objectives. The first significant event was Obama's failure to respond to Assad's use of chemical weapons in Eastern Ghouta in 2013. The US' lack of assertive response despite its threats to act if this red line was crossed, has had enormous ramifications for the role of external actors within Syria for the duration of the war. Indeed, Obama's decision was perceived as a strong signal that even in the face of chemical weapons, the UK and US would not intervene against Assad.

The second event was the rise of Islamic State and the declaration of a new Caliphate on 29 June 2014. The organisation not only threatened the US' interests in Iraq, but it also carried out murderous attacks in European cities and on American soil (Hashim 2014). In response, Washington, under Obama's administration established an international coalition to combat the group, signalling a shift from the previous policy in support of regime change in Damascus and focusing solely on this new objective to push back Islamic State (Kumar 2015). This recalibration of priorities led to the development of two US policies. First, a tolerance towards a proliferation of Iranian-backed Shia armed groups in conflict with Islamic State, thus angering the Arab Gulf monarchies (Mansour 2017). Second, a train-and-equip programme to Free Syrian Army (FSA) forces and the People Protection Units (YPG), the armed wing of the Syrian Kurdistan Workers' Party (PKK) affiliate the Democratic Union Party (PYD) (Parlar Dal 2016). This alienated Turkey who lists both groups as terrorist organisations (Ertem 2018).

The third significant event that changed the course of the Syrian conflict was the Egyptian coup-d'état on 3 July 2013. Even though not directly related to Syria, this event signalled the start of a Saudi-Emirati-led counter-revolution in the Middle East (Steinberg 2014). The Arab Gulf monarchies exploited the Syrian conflict to counter Iran, but they were also very wary of the Arab Spring and the wave of democratisation it promised for the region. These fears increased as Turkey and Qatar seemed more in tune with the revolutionary movements, thus threatening Riyadh's leadership of the Arab Sunni states. In Syria, Saudi Arabia and the United Arab Emirates anticipated a possible opposition victory and directed their support to amenable groups polarising the opposition and weakening its unified stance against Assad.

These shifts coincided with Washington's efforts to negotiate the Joint Comprehensive Plan of Action (JCPOA) with Tehran, which further exacerbated the feelings of betrayal within the Gulf Cooperation Council (GCC), who perceived the agreement as a green light for Iran's expansion in

Syria and the Middle East (Bahi 2017). Moreover, feeling the US reluctance to support the opposition against Assad or Iran in Syria, Russia officially announced its direct military intervention in the Syrian conflict in September 2015, effectively ending all prospects of Western-led efforts to oust Assad. The US-GCC relations were rectified later on when Trump announced the US withdrawal from the JCPOA. This resolved perceived grievances between Riyadh and Abu Dhabi. But it fell short of recalibrating the balance of power between the opposition and the Syrian regime, which now felt stronger because of the air superiority the Russian forces brought to the battlefield.

The US foreign policy changes in the Middles East under the new administration did not have a similar impact on Turkey. The latter fear of YPG expansion in Northern Syria has led its leadership to seek a security arrangement with Russia. This unusual cooperation allowed the Turkish Army with the help and assistance of opposition armed groups to chase the YPG out of northern Aleppo but has also further strained Ankara's relationship with the rest of the NATO nations (Kasapoğlu 2018).

Model of collaboration and cooperation

The US, Russia, Iran, Turkey, France and the UK all have boots on the ground, but they are present in small numbers, rarely fight on the frontline and mostly provide technical and logistic support to their Syrian allies. The models of their interaction and collaboration with the latter differ from one case to another. The Russians, for instance, adopt a top to bottom approach in dealing with loyalist forces, enforcing direct oversight of the regime's military operations, and even intervening in nominating and promoting their commanders (Al-Modon 2019). Others, like the US, adopt a bottom-up approach, assisting grass-root movements without significant interference in their clients' organisation or modus operandi beyond vetting eligible members for training or receiving funds and equipment.

Generally, differences among these models can be observed over three main spectrums. First, the degree of ideological alignment and extent to which the foreign power requires a certain level of affinity with the local ally. Second, the level of professionalism expected from the armed group or army brigade. Third, the modalities of support provided to the partner and to how it is put to his disposal.

Naturally, global and extra-regional powers score low on the ideological alignment spectrum. The lack of religious, sectarian and cultural similarities with the local communities and groups does not allow them to impose high standards in picking their allies. Alternatively, they choose to affiliate groups depending on their differences with their adversaries. Russia, for instance, initially supported hardcore loyalists but also accommodated former reconciled opposition fighters (al-Khateb 2019). Similarly, the US first assisted radical Sunni Arab opposition groups but then turned to the leftist/communist YPG when the former refused to put on hold their battles against Assad forces to concentrate on the emerging threat posed by Islamic State (Blanchard 2014; Kanat 2015). In contrast, regional actors relied on groups with higher sectarian or ideological affinity. For instance, Turkey, Saudi Arabia and Qatar sought Sunni-dominated groups, whereas Iran mostly supported Shia or Alawites. For example, when the Turkish army launched the Olive Branch operation against the YPG in Afrin, it relied on Turkmen and Arabs with grievances against the Kurds (Etgu 2018). However, it is worth noting that Iran displayed an advanced capacity to recruit beyond the traditional sectarian rifts and was able to ideologically indoctrinate local leaders, including Sunnis, to compensate for the relatively small Shia community in Syria.

As for the level of professionalism, the US and Russia stand on opposing ends of the spectrum. The US, on the one hand, has shown little interest in how their allies are organised, and this trait has been witnessed during their interaction with the Arab Sunni opposition as well as the Kurds. Only a few attempts to form standard military operations rooms were observed throughout the American intervention in Syria (Lister 2016). Russia, on the other hand, has early on demonstrated a strong desire to reform the Syrian state military and security apparatus. This determination could be explained by the Russian intent to instate a disciplined satellite state in the Levant, an objective the US never had. Like Russia, Turkey also has adopted a similar aim of professionalising the opposition groups in northern Syria. This is motivated by Ankara's plans to endure a continued presence in Syria, but it is also due to the Turkish army's lack of experience working with non-state actors.

In contrast, Iran and the GCC states both have histories of operating with grassroots movements and militias. Iran has not only shipped Lebanese, Iraqi, Afghani, and Pakistani militants to Syria but has encouraged the formation of Syrian loyalist groups (Mohseni 2017). Likewise, the Arab Gulf states also facilitated the flow of hundreds of volunteers to join the opposition and poured in money and equipment to a myriad of Sunni Arab armed groups in the early stages of the conflict (Hokayem, 2014).

In regard to the support modalities provided to their allies, the regime backers have proven more generous and direct. Reports suggest that Assad requested Iranian technical assistance as soon as 2011, mainly in an advisory capacity to train the regime forces in containing protests. But this assistance has also involved financial aid to the Syrian government (Fulton 2013). In 2013, the concerted Iranian efforts to preserve Bashar al-Assad in power significantly increased following the rapid advance of the Syrian opposition in northern and central Syria, and hundreds of Islamic Revolutionary Guard Corps experts and equipment were put forward to prevent the opposition's victory. This level of support was only matched by the Russians in September 2015, when their aviation and military experts stepped in to compensate for the losses the Syrian Army and Iranian-backed militias endured, and to reverse the opposition's military gains. In comparison, the opposition allies did not invest nearly as much. The US mainly provided small arms ammunition, short-ranged artilleries, limited amounts of anti-tank guided missiles, and coordinated air-strikes against Islamic State militants only (Krieg 2016). Moreover, the Arab Gulf states and Turkey took a similar approach by not offering more than limited financial and logistical support to their affiliated groups.

Seeking autonomy

Foreign powers have undoubtedly influenced the course of the conflict and have guaranteed a say for themselves in its final resolution. However, local groups and actors have still managed to preserve a certain level of agency and have, on multiple occasions, impacted their backers' policies. Engaged in remote warfare, foreign backers can only exercise a limited degree of control and restraint over their local allies. Hence, it is only natural that the latter pursue their policies, especially in their domestic sphere of influence and mainly in security, economy and political engagement with their constituencies. Moreover, as the conflict persists, these actors embrace an increasing level of autonomy, and heterogeneous governance structures emerge, adding complexity to future peace resolution as local communities become more protective of their new acquisitions and rights.

From a practical point of view, Syrian armed groups are key actors in security provision, they are de-facto governments within the territories under their control, they are military entities active in combat, and they behave as authorities responsible for the protection of their constituencies. To some extent, all foreign backers seek stabilisation in their respective sphere of influence, but their level of engagement differs from one another. On the one hand, the US encourage their allies to embrace inclusive policies and to recruit from other ethnic and religious communities to lessen grievances among minorities. Turkey and Russia, on the other hand, pursue a more direct approach and often discipline rogue actors if proven guilty or a threat to their stabilisation efforts. Nonetheless, Syrian groups still assume a significant autonomous role in security provision. For instance, the PYD created a sperate unit for law reinforcement in their areas of control called the Asayish (Federici 2015). Whereas the Coalition collaborates with and coordinates the YPG on the battlefield, the Asayish remains autonomous as they pursue their policies independently from any foreign intervention. Similarly, the opposition armed groups are only partially involved in the day-to-day disputes, and they often resort to local mediation and arbitration in maintaining societal peace. This approach prevents unwelcomed interference in domestic politics (Hatahet 2017).

Economically speaking, the principle that combatants should be separated from civilians often makes little sense to the armed groups. On the contrary, they rely heavily on their proximity to civilian populations to sustain themselves and to consolidate their control over a territory and its resources. Moreover, combatants often engage in parallel financial activities, trade, trafficking, smuggling, extortion, and various quick on cash activities. It is true that the majority of Syrian armed groups heavily rely on their backers' financial aid and assistance, but they also grew accustomed to developing their own sources of income. For instance, the YPG weapons and equipment is mostly provided by the US (Ergun 2018). Nonetheless, they also control a large section of the oil production in Syria, offering them alternative revenue streams and consequently more autonomy. Indeed, multiple reports confirm active oil trade between the regime and YPG despite the US sanctions placed over Damascus (Benoit Faucon 2019). Likewise, the opposition and loyalist groups also engage in bilateral trade and cooperate in illegal smuggling activities.

Lastly, Syrian armed groups are sophisticated entities that seek political legitimacy within their constituency. To enhance their political stance, the majority of non-state actors sought to provide humanitarian and social services for their people. Both the YPG and Syrian opposition established local councils to provide basic governance structures and to represent their constituencies. The regime, on the other, has tolerated the emergence of the popular informal committee to manage the daily governance aspect of loyalist communities (Agha 2019). Here again, foreign backers mostly dealt with these grass root institutions as a reality and were not able to command and control them beyond their capacity to fund their activities and with little success when this occurred.

Conclusion

Three main broad camps currently exist in Syria. The first is led and funded by the US and is composed of the YPG, local Arab tribes and Assyrian armed groups. The second is backed by Russia and Iran and consists of regime forces, local militias, and foreign Shia militant groups. The third is endorsed by Turkey and incorporates a myriad of opposition groups from various ideological stances, including mainstream Islamist factions and national patriotic parties. Even though all foreign backers are engaged in different shapes and forms of negotiations to end the stalemate and establish a peaceful resolution as soon as possible, each camp is fundamentally at odds with the other.

The Russian intervention in September 2015 marked a turning point in the Syrian conflict and has theoretically ended all possibilities of foreign involvement that achieves a regime change. Moscow's commitment to Assad has placed him in a stronger position vis-à-vis his opponents. The Syrian regime's current objective is to regain control over all the Syrian territory and to reconsolidate its authority over all armed groups, including loyalists and Iranian-backed local allies. The YPG thus far has restrained from any confrontation with the regime but is nowhere ready to abandon its autonomy, which puts it on a collision course with Damascus. The Peace Spring military operation launched by Turkey on north-eastern Syria, however, inadvertently unlocked the deadlock on SDF-Assad negotiations. Unable to obtain the level of protection it required from Washington, the Autonomous Administration of North and East Syria (commonly known as AANES) sought Moscow and Damascus to deter Ankara and has concluded a hasty security arrangement with the regime that allows the latter to redeploy its forces along the Syrian borders with Turkey. In a sense, such an arrangement also plays in favour of Turkey, who believes it could obtain a stronger commitment from Russia to contain the PYD than the US, but most importantly it also sets the environment for the second round of political negotiations between the SDF and Assad.

It is still too early to draw the potential outcome of such negotiations, especially when taking into consideration the US decision to remain in the area. However, in contrast to its approach to dealing with Turkey-backed Arab armed opposition groups, the regime has not dismissed the possibility of a political arrangement with the PYD. The latter's dependence on Russia's protecting it has considerably weakened its stance towards Damascus. The remaining question is whether Assad would conclude such an accord 'domestically' or would allow it to become an internationally led process.

Russia has been relatively more successful in its remote warfare in Syria than other actors, including the US, for several reasons. First, Moscow's stance has been consistent throughout the conflict; it has demonstrated unmatched willingness in supporting Assad. In comparison, both Washington and regional backers of the opposition have considerably changed their position. The US' initial backing of the Syrian rebels was done to control the flow of weapons and funds to them rather than actively seeking the regime change. The US has since abandoned all resolve to support them as Islamic State emerged. Similarly, actively engaged in assisting and equipping Assad opponents, Turkey eventually shifted all attention to countering the US empowered PYD as their influence grew at its southern borders, and the oil-rich Arab monarchies actively withdrew from Syria following the Houthis' 2015 take-over of Sana'a in Yemen.

Second, Moscow has also shown more strategic agility than Washington. Putin took advantage of the domestic turmoil that shook Turkey to neutralise its stance towards Assad. Moreover, when given the opportunity, he has exhibited more sensitivity to Ankara's security concern in Syria and has successfully avoided raising any significant contention with other regional powers. In comparison, the US has shown less resolve to address its Turkish ally's fears, and the Coalition's support to the PYD increased its anxieties and has indirectly pushed it into Russia's arms.

Third, protected by its veto in the Security Council and facing less domestic scrutiny over its use of military force abroad, Russia has shown no constraint in defeating its opponents. Implementing scorched-earth tactics, the Russian air forces relentlessly destroyed all opposition capacity to resist and have also seized the opportunity to test new weapons. The US, on the other hand, has demonstrated this willingness only against Islamic State, and even then, it has refrained from using excessive force. The Russian inclination to use all necessary means to claim victory has instilled more fear in its adversaries, and its threats were thus taken more seriously than the less assertive US in Syria.

Overall, the conflict in Syria presents an interesting example of modern conflicts, with global and regional powers waging remote warfare against their adversaries. In comparison with other contemporary wars, foreign backers have pursued their objectives with minimum human and capital costs. However, the Syrian case is also an excellent illustration of the limits of remote warfare. Local armed and political groups are gaining maturity and are increasingly imposing their footprint on the regional scene. Meanwhile, the centralised nation-state model of governance is eroding, and no credible structures are emerging to fill the void left by the collapse of authoritarian regimes. Hence, it is crucial to recognise the need for establishing a new political framework to build sustainable peace in the Middle East. Such a structure should place community participation and consensus at the heart of any political process. Otherwise, the region will remain a hotbed for insurgency and instability for years to come.

References

Achen, Christopher H., and Duncan Snidal. 1989. 'Rational Deterrence Theory and Comparative Case Studies.' *World Politics,* 41(2): 144–169.

Agha, Munqeth Othman. 2019. 'Early Recovery in Syria: An Assessment of the Regime's Role and Capability.' In *Economic Recovery in Syria: Mapping Actors and Assessing Current Policies*, 84. Omran for Strategic Studies.

al-Khateb, Khaled. 2019. 'Ex-FSA fighters recruited by Damascus to fight opposition in northern Syria.' *Al-Monitor*. 13 June. https://www.al-monitor.com/pulse/originals/2019/06/syrian-regime-recruit-former-fsa-fighters-idlib-offensive.html

Al-Modon. 2019. 'Russia and Iran Divide up Syria's Security.' *The Syrian Observer.* 8 April. https://syrianobserver.com/EN/features/49638/russia-and-iran-divide-up-syrias-security.html

Etgu, Muslum. 2018. 'Syrian Arab and Turkmen tribes hail Turkish operation.' Anadolu Agency.13 December. https://www.aa.com.tr/en/middle-east/syrian-arab-and-turkmen-tribes-hail-turkish-operation/1337657

Bahi, Riham. 2017. 'Iran, the GCC and the Implications of the Nuclear Deal: Rivalry versus Engagement.' *The International Spectator,* 52(2): 89–101.

Benoit, Faucon, and Nazih Osseiran. 2019. 'US's Syria Ally Supplies Oil to Assad's Brokers.' *Wall Street Journal.* 8 February. https://www.wsj.com/articles/u-s-s-syria-ally-supplies-oil-to-assads-brokers-11549645073

Blanchard, Christopher M., Carla E. Humud, and Mary Beth D. Nikitin. 2014. *Armed conflict in Syria: Overview and US response.* Library of Congress Washington DC Congressional Research Service.

Bojicic-Dzelilovic, Vesna, and Rim Turkmani. 2018. 'War economy, governance and security in Syria's opposition-controlled areas.' *Stability: International Journal of Security and Development,* 7(1): 1–17.

Ergun, Doruk. 2018. 'External Actors and VNSAs: An Analysis of the United States, Russia, ISIS, and PYD/YPG.' In *Violent Non-state Actors and the Syrian Civil War,* edited by Ozden Zeynep Oktav, Emel Parlar Dal and Ali Murat Kursun. Springer: 149–172.

Ertem, Helin Sarı. 2018. 'Surrogate Warfare in Syria and the Pitfalls of Diverging US Attitudes Toward ISIS and PYD/YPG. In *Violent Non-state Actors and the Syrian Civil War,* edited by Ozden Zeynep Oktav, Emel Parlar Dal and Ali Murat Kursun. Springer: 129–148.

Federici, Vittoria. 2015. 'The rise of Rojava: Kurdish autonomy in the Syrian conflict.' *SAIS Review of International Affairs,* 35(2): 81–90.

Fulton, Will et al. 2013. *Iranian Strategy in Syria.* Institute for the Study of War.

Hashim, Ahmed S. 2014. 'The Islamic State: From al-Qaeda Affiliate to Caliphate.' *Middle East Policy* 21(4): 69–83.

Hatahet, Sinan. 2017. 'Infighting Among the Syrian Opposition.' *Sharq Forum.*

Hinnebusch, Raymond. 2017. 'The battle for Syria: international rivalry in the new Middle East.' *International Affairs,* 93(1): 223–224.

Hokayem, Emile. 2014. 'Iran, the Gulf States and the Syrian civil war.' S*urvival,* 56(6*):* 59– 86.

Kagan, Frederick W., et al. 2016. *Al-Qaeda and ISIS: Existential Threats to the US and Europe.* Institute for the Study of War.

Kanat, Kilic, and Kadir Ustun. 2015. 'US-Turkey Realignment on Syria.' *Middle East Policy,* 22(4): 88–97.

Kasapoğlu, Can, and Sinan Ülgen. 2018. 'Operation Olive Branch: A Political–Military Assessment.' Centre for Economics and Foreign Policy Studies (EDAM).

Katz, Mark, and Va Fairfax. 2011. 'Russia and the Arab spring.' *Russian Analytical Digest* 98(6): 4–6.

Khalaf, Rana. 2015. 'Governance without Government in Syria: Civil society and state building during conflict.' *Syria Studies.* 7(3).

Krieg, Andreas. 2016. 'Externalizing the burden of war: the Obama Doctrine and US foreign policy in the Middle East.' *International Affairs,* 92(1): 97–113.

Kumar, Chanchal. 2015. 'Islamic State of Iraq and Syria (ISIS) a global threat: international strategy to counter the threat.' *Journal of Social Sciences and Humanities,* 1(4): 345–353.

Lister, Charles. 2016. 'The Free Syrian Army: A decentralized insurgent brand.' The Brookings Project on US Relations with the Islamic World. Brookings Institute.

Mansour, Renad. 2017. 'Iraq After the Fall of ISIS: The Struggle for the State.' Middle East and North Africa Programme, Chatham House.

Mohseni, Payam, and Hussein Kalout. 2017. 'Iran's axis of resistance rises.' *Foreign Affairs.* 24 January.

Pala, Özgür, and Bülent Aras. 2015. 'Practical geopolitical reasoning in the Turkish and Qatari foreign policy on the Arab Spring.' *Journal of Balkan and Near Eastern Studies,* 17(3): 286–302.

Parlar Dal, Emel. 2016. 'Impact of the transnationalization of the Syrian civil war on Turkey: conflict spillover cases of ISIS and PYD-YPG/PKK.' *Cambridge Review of International Affairs,* 29(4): 1396–1420.

Phillips, Christopher. 2015. 'Sectarianism and conflict in Syria.' *Third World Quarterly,* 36(2): 357–376.

Rabinovich, Itamar. 2017. 'The Syrian Civil War as a Global Crisis.' *Sfera Politicii,* 25 (191-192): 44–48.

Rich, Ben, and Dara Conduit. 2015. 'The impact of jihadist foreign fighters on indigenous secular-nationalist causes: Contrasting Chechnya and Syria.' *Studies in Conflict and Terrorism,* 38(2): 113– 131.

Steinberg, Guido. 2014. 'Leading the counter-revolution: Saudi Arabia and the Arab Spring.' *Social Science Open Access Repository* (SWP) 27.

11

Death by Data: Drones, Kill Lists and Algorithms

JENNIFER GIBSON

In 2018 Google employees made headlines when they openly protested the company's involvement in Project Maven – a controversial US programme aimed at integrating artificial intelligence into military operations. Google argued it was simply helping automate analysis of drone footage. Employees signed an open letter to CEO Sundar Pichai arguing Google 'should not be in the business of war' (BBC 2018). For many communities in places like Pakistan and Yemen, computers are already making life and death decisions. Massive amounts of signals intelligence are being run through algorithms that make decisions as to who is 'suspicious' and who 'isn't.' For populations with a drone flying overhead, those decisions can be deadly. Nobody knows the damage America's covert drone war can wreak better than Faisal bin ali Jaber. Faisal's brother-in-law, Salem, was killed by a drone just days after he preached against al-Qaeda in 2012 (Jaber 2016). The strike was likely a 'signature' strike, one taken based on a suspicious 'pattern of behaviour.' This chapter will examine the case of Faisal bin ali Jaber and address just some of the troubling questions that arise as big data and remote warfare converge. Can targeting based on metadata ever be compliant with international humanitarian law (IHL) and its principle of 'distinction' and just what are the 'feasible' precautions the US must take to ensure it is?

America's drone wars: the case of Faisal bin Ali Jaber

While the first known drone strike took place in Afghanistan in 2001, it was not until President Obama came into office that drones became the weapon of choice in the United States' 'War on Terror'. Dubbed by some 'The Drone Presidency', President Obama used drones to carry out at least 563 strikes during his time in office, ten times more than his predecessor George W.

Bush (Cole 2016). Controversially, the strikes were taken outside of traditional battlefields in places like Yemen, Pakistan and Somalia, and killed potentially as many as 4936 people (Purkiss and Serle 2017).

One of the biggest criticisms levelled at the programme – both under President Obama and now under President Trump – is the degree of secrecy with which it is carried out and the lack of accountability for mistakes that have been made. President Obama failed to even acknowledge the programme's existence until early in his second term (Obama 2013), and it took another three years before his administration would release the first accounting of civilian harm (Shane 2016). That accounting estimated that almost eight years of strikes had killed between 64 and 117 'non-combatants', a range significantly lower than independent estimates, which ranged from 207 (excluding Somalia) to 801 (Shane 2016). The *New York Times*' Scott Shane wrote that '[i]t showed that even inside the government, there is no certainty about whom it has killed' (Shane 2016).

The data when it was released also failed to include very basic details, such as when and where those civilian casualties occurred (Zenko 2016). This made it impossible for human rights groups and independent monitors to compare their own numbers to the government's figures or to assess why there were such wide discrepancies. It also left the families of those who lost loved ones asking: 'Has my family been counted?' (Jaber 2016).

Faisal bin ali Jaber was one of those people asking. An engineer from Yemen, Faisal's brother-in-law, Salem, and nephew, Waleed, were killed by a US drone strike on 29 August 2012. Salem was an imam who was known for speaking out against al-Qaeda in his sermons, and Waleed was one of only two policemen in their local village of Khashamir. The Friday before he was killed, Salem gave a sermon at the mosque, denouncing al-Qaeda's ideology. The sermon so strongly denounced al-Qaeda that Faisal would later state that members of the family were worried he might be in danger of reprisals from the group. When Faisal spoke to Salem about the family's concerns, Salem responded: 'If I don't use my position to make it clear to my congregation that this ideology is wrong, who will? I will die anyway, and I would rather die saying what I believe than die silent' (Jaber 2013).

Shortly after the sermon, three young men arrived in the village, demanding to speak with Salem. Worried about security and concerned they might be al-Qaeda, Salem eventually agreed to meet them, but took Waleed with him for protection. They agreed to meet outside the mosque in the open, where Salem and Waleed thought it would be safest. Within minutes of stepping out of the mosque to meet the three young men, a drone hovering overhead fired, killing all five people (Ibid.).

Faisal was present on the day Salem was killed, as the entire family had gathered to attend his eldest son's wedding. Instead of celebrating, they spent the day collecting body parts. When the Yemeni security services arrived an hour after the strike, Faisal asked them why they waited to strike until Salem and Waleed were present. There was a checkpoint less than 1km from the village that the men must have travelled through to reach the village and a military base 3km away. They had no answers (Ibid.).

Faisal went looking for answers and in November 2013, he travelled to the US to speak to Congress and meet with officials from the National Security Agency. The headline on the front page of the *New York Times* summed up the trip: 'Questions on Drone Strike Find Only Silence' (Shane 2013). Eight months later, one of Faisal's relatives was offered a bag containing $100,000 in sequentially marked US dollar bills at a meeting with the Yemeni National Security Bureau (NSB). The NSB official told a family representative that the money was from the US and that he had been asked to pass it along. When the family asked for official written notification of who it was from, the security agents refused (Isikoff 2014).

In 2015, Faisal filed a civil claim against the US Government seeking an apology and a declaration that the strike which killed his relatives was unlawful. He did not seek compensation, instead asking only for the acknowledgment that did not come with the cash secretly offered to his family in 2014. In the suit, Faisal also questioned why, in the apparent absence of any immediate threat, the three unidentified targets could not have been detained safely by Yemeni forces at checkpoints or, failing that, why the missiles could not have been fired sooner when the targets were isolated (Ahmed Salem Bin Ali Jaber v United States 2015).

Despite new information showing that US officials knew shortly after the strike that Salem and Waleed were civilians (Currier, Devereaux and Scahill 2015), in June 2017 the Court of Appeal for the District of Columbia rejected Faisal's case. The Court ruled unanimously that it could not rule on the matter, citing precedent preventing the judicial branch from adjudicating 'political questions.' However, in a rare concurring opinion, Judge Janice Rogers Brown issued an unprecedented rebuke of the drone programme. She pronounced American democracy 'broken' and congressional oversight a 'joke' in failing to check the US drone killing programme. The judge, an appointee of President Obama's predecessor, George W. Bush, seemed troubled that current legal precedent prevented her court from acting as a check on potential executive war crimes. Calling drone strikes 'outsized power', she questioned who would be left to keep them in 'check' stating that 'it is up to others to take it from here' (Ahmed Salem Bin Ali Jaber v United States 2017).

The role of algorithms in US targeted killings

Faisal has never received answers to why, or how, his family was targeted. It was likely though a 'signature' strike gone wrong. The precision language that surrounds drones and the US targeted killing programme suggests that the US always knows the identity of the individuals it targets. The reality is much different.

There are two main types of strikes: 'personality' strikes taken against known, named high value targets, and 'signature' strikes, which are taken based upon 'suspicious' patterns of behaviour. (Becker and Shane 2012) President Obama authorised both types of strikes in Yemen and Pakistan, with the criteria for taking such strikes widely regarded as too lax. One intelligence agent, speaking anonymously to the *New York Times*, said the 'joke' was that the CIA 'sees three guys doing jumping jacks' and the agency suspects a terrorist training camp (Ibid.).

The practice elicited widespread criticism, with a variety of actors raising concerns about the legality of such strikes, the civilian harm they engendered, and the potential counter-productivity of killing individuals you could not even identify. In response, President Obama signalled in May 2013 that the US would take steps to phase out this controversial tactic (New York Times 2013). His successor, President Trump, reportedly reinstated them within months of coming into office (Dilanian, Nichols and Kube 2017).

What has become clear through leaks to the media is that the 'signature' upon which both administrations relied is far less visual and far more data driven. Lethal drone strikes are the culmination of a complex process that involves the collection of data and intelligence through mass surveillance programmes that hoover up millions of calls, emails and other means of electronic communications. Surveillance drones gather countless images and videos which are analysed and fed into the identification and location of suspects. In April 2014, General Michael Hayden, former Director of the CIA, told a John Hopkins University Symposium that the United States 'kills people based on metadata' (Cole 2014).

This is especially true in places like Yemen, where the US has a limited footprint. Without human sources on the ground, it is overly reliant on signals intelligence from computers and cell phones, and the quality of those intercepts is limited (Currier and Maass 2015). Moreover, according to a leaked US military document in 2013, signals intelligence is often supplied by foreign governments with their own agendas (Ibid.). Such questionable signals intelligence makes up more than half the intelligence collected on targets (Ibid.).

The remaining signals intelligence is collected through mass surveillance programmes run by the United States and its European allies, including the UK. Through classified programmes, such as OVERHEAD, GHOST HUNTER and APPARITION, the US and its allies have been hoovering up intelligence from satellites, radio and cell phone towers in countries like Yemen and Pakistan for the express purpose of identifying and locating targets (Gallagher 2016). The aim of such programmes, according to one leaked document about a programme called GHOSTWOLF, is to 'support efforts to capture or eliminate key nodes in terrorist networks' (GCHQ 2011).

According to one drone operator, the United States often locates drone targets by analysing the activity of a SIM card, rather than the actual content of the calls. He said the problem with this is that they frequently have no idea who is holding the cell phone they target:

> 'People get hung up that there's a targeted list of people. It's really like we're targeting a cell phone. We're not going after people – we're going after their phones, in the hopes that the person on the other end of that missile is the bad guy.'

> 'Once the bomb lands…you the know the phone is there. But we don't know who's behind it, who's holding it. It's of course assumed that the phone belongs to a human being who is nefarious and considered an 'unlawful enemy combatant. This is where it gets very shady.' (quoted in Scahill and Greenwald 2014)

A leaked document from Edward Snowden shows just how this data is then fed into algorithms that help the US identify targets. According to a document titled 'SKYNET: Applying Advanced Cloud-based Behavior Analytics', the US developed a programme called SKYNET that it used to identify suspected terrorists based on their metadata – the electronic patterns of their communications, writings, social media postings and travel. According to one slide, SKYNET 'applies complex combinations of geospatial, geotemporal, pattern-of-life and travel analytics to bulk DNR [Dial Number Recognition] data to identify patterns of suspect activity.' Put more plainly, the programme used 'behaviour-based analytics' to run data such as travel patterns, 'frequent handset swapping or powering down', low-use phone activity, or frequent disconnections from the phone network through an algorithm which then identified those who fit the 'pattern' of a terrorist (National Security Agency 2015).

The essential flaw in the use of automated algorithms to select targets is aptly demonstrated by the individual SKYNET itself identified – Ahmed Zaidan. Ahmed Zaidan, the former Bureau Chief of Al Jazeera in Pakistan, is a ground-breaking journalist who managed to interview Osama bin Laden twice before September 2001. As part of his job in Pakistan, he regularly interviewed those associated with al-Qaeda and other militant groups. Yet SKYNET still classified him as the 'highest scoring' target, identifying him as a courier for al-Qaeda in part because of his travel patterns (National Security Agency 2015).

The use of metadata in targeting – IHL compliant?

Zaidan's case aptly demonstrates the problematic nature of targeting based on algorithms and raises questions about just how certain metadata can be. One of the foundational principles of international humanitarian law[1] is the protection of civilians in conflict. In order to ensure such protection, Article 51(2) of the First Additional Protocol to the Geneva Conventions, requires parties to a conflict to 'distinguish' between combatants and non-combatants when targeting individuals for lethal force (Henckaerts and Doswald-Beck 2009, 3). This principle applies to both international and non-international armed conflicts and has been described by the International Court of Justice (ICJ) as the 'cardinal rule' of IHL (Corn 2011–12, 441).

Because the principle of distinction is so central to IHL, Article 57(2) of the First Additional Protocol states that it is essential that those who are planning to carry out the attack should '[d]o everything feasible to verify that the objectives to be attacked are neither civilians nor civilian objects.' (Henckaerts and Doswald-Beck 2009). In situations where there is still 'doubt' that an individual is a legitimate target after taking all 'feasible' precautions, Article 50(1) states 'that person shall be considered to be a civilian.' (Henckaerts and Doswald-Beck 2009).

The question that therefore arises in the context of the use of metadata and targeting is how good is metadata at distinguishing between civilians and those 'directly participating in hostilities' or carrying out a 'continuous combat

---

[1]   There is significant controversy over whether US drone strikes in places like Pakistan and Yemen are indeed part of an armed conflict. There is not scope in this paper to go into depth on this debate, and so for the purposes of argument it assumes that they are. This is because if they are not, international human rights law, not international humanitarian law, would apply. The former is a much stricter legal standard on the use of lethal force and if a strike, or method, does not meet the bare requirements of international humanitarian law, it will in all cases fail to meet those set out by international human rights law.

function'? And is that sufficient enough to meet the 'feasibility' standard set out by the International Committee of the Red Cross, which requires a party to a conflict to take 'those precautions which are practically possible taking into account all the other circumstances' (Rogers 2016).

The evidence to date would suggest it is not. Take for instance the SKYNET programme leaked by Edward Snowden. While we do not know whether the United States ever used the programme to carry out lethal action, we do know based on Michael Hayden's comments that it likely used some form of algorithm to 'kill people based on metadata.' SKYNET therefore provides a window into the type of programmes the US is developing and the types of problems that might arise.

Patrick Ball, a data scientist and the Director of Research at the Human Rights Data Analysis Group, has previously given expert testimony before war crimes tribunals. After reviewing the SKYNET slides, he identified several flaws in the way the algorithm worked, which made the results scientifically unsound. One of the key flaws was that there are *very few* 'known terrorists' for the National Security Agency (NSA) to use to train *and test* the model. A typical approach to testing the model would be to give it records it has never seen, but according to Ball, if the NSA is using the same profiles the model has already seen then 'their classification fit assessment [will be] ridiculously optimistic.' He goes on to point out that a false positive rate of 0.008 would be remarkably low if this were being used in a business context, but when applied to the general Pakistani population, it means 15,000 people would be misclassified as 'terrorists' (Grothoff and Porup 2016).

Security expert Bruce Schneier agrees with Patrick Ball:

> Government uses of big data are inherently different from corporate uses. The accuracy requirements mean that the same technology doesn't work. If Google makes a mistake, people see an ad for a car they don't want to buy. If the government makes a mistake, they kill innocents. (Grothoff and Porup 2016)

Looking beyond the programme itself, there is also the evidence emanating from on the ground investigations by independent organisations. Take, for example, independent monitoring by the Bureau of Investigative Journalism, which suggests that as many as a quarter of those killed under the Obama administration may have been civilians (Purkiss and Serle 2017). Experts believe that 'signature' strikes, such as the one that killed Faisal's family members, likely accounted for the vast majority of these (Dworkin 2013).

In March of this year, a German court questioned the lawfulness of US strikes in Yemen in part because of their impact on civilians. The case was brought by Faisal in 2014, arguing that the continued use of Ramstein Air Base by the US for strikes in Yemen threatened his and his family's right to life under the German constitution. In March 2019, the German high court agreed. It found that 'at least part of the US armed drone strikes [...] in Yemen are not compatible with international law', and that Germany must do more to ensure its territory is not used to carry out unlawful strikes (Bin Ali Jaber v Germany 2019). In its decision, the court acknowledged that Faisal and his family 'are justified in fearing risks to life and limb from US drone strikes that use Ramstein Air base in violation of International Law' (Bin Ali Jaber v Germany 2019).

A factor in the court's decision was significant increase in drone strikes since President Donald Trump took office, and his administration's reported rollback of safeguards that were intended to protect civilians, including the renewed use of signature strikes. The court went on to state that there were 'weighty indicators to suggest that at least part of the US armed drone strikes…in Yemen are not compatible with international law and that plaintiffs' right to life is therefore unlawfully compromised.' (Bin Ali Jaber v Germany 2019) The German Government's declarations to the contrary, according to the court, were based on 'insufficient fact-finding and ultimately not legally sustainable' (Bin Ali Jaber v Germany 2019). The court also noted that the fact that Faisal and his family were denied a judicial review by the American courts of their relatives' deaths 'runs counter to the idea that there were any [independent investigations by US authorities' (Bin Ali Jaber v Germany 2019).

Conclusion: 'feasible' precautions and the death of Faisal's family

The German court's decision highlights a key facet of the legal question surrounding the use of metadata in targeting: that adequate, independent post-strike investigations are the bare minimum of what 'feasible' precautions should include. Algorithms, at their best, merely tell us about relationships. They don't tell us whether Faisal's brother-in-law is meeting with three young men because he's planning an attack with them, or instead, if the meeting is to explain why he believes al-Qaeda's ideology is wrong. Metadata cannot tell us whether Ahmed Zaidan is meeting with known fighters because he too plans to fight, or because he is a journalist doing his job. Moreover, even the best algorithms only work after constant testing and refinement, the type of refinement that requires one to identify and correct for errors. Without post-strike investigations, there is no way the US can tell whether their strikes have hit lawful targets. Without this information, there is no way they can take 'feasible' precautions to ensure the mistakes like those that killed Faisal's family do not happen again.

References

Ahmed Salem Bin Ali Jaber v United States. 2017. 861 F.3d 241 (DC Circuit, 30 June 2017).

Ahmed Salem Bin Ali Jaber v United States. 2015. 1:15-cv-00840 (DC District Court, 8 June 2015).

BBC. 2019. 'Google should not be in business of war, say employees.' 5 April 2018. Accessed 11 July 2019. https://www.bbc.co.uk/news/business-43656378

Becker, Jo, and Scott Shane. 2012. 'Secret 'Kill List' Tests Obama's Principles.' *New York Times*. 29 May.

Bin Ali Jaber v Germany. 4 A 1361/15 (North Rhine-Westphalia Higher Administrative Court, 19 March 2019).

Cole, David. 2014. 'We Kill People Based on Metadata.' *New York Review of Books*. 10 May

————. 2016. 'The Drone Presidency.' *The New York Review of Books,* 13th edition. 18 August.

Corn, Gregory. 2012. 'Targeting, Command Judgment, and a Proposed Quantum of Information Component.' *Brooklyn Law Review*, 2011-12: 437–498.

Currier, Cora, and Peter Maass. 2015. *'Firing Blind: Flawed Intelligence and the Limits of Drone Technology.' The Intercept*. 15 October. Accessed 14 July 2019. https://theintercept.com/drone-papers/firing-blind/

Currier, Cora, Ryan Devereaux and Jeremy Scahill. 2015. 'Secret Details of Drone Strike Revealed as Unprecedented Case Goes to German Court.' 17 April. Accessed 19 May 2019. https://theintercept.com/2015/04/17/victim-u-s-drone-strike-gets-day-german-court/

Dilanian, Ken, Hans Nichols and Courtney Kube. 2017. 'Trump Admin Ups Drone Strikes, Tolerates More Civilian Deaths: US Officials.' *NBC News*. March 14. Accessed 15 July 2019. https://www.nbcnews.com/news/us-news/trump-admin-ups-drone-strikes-tolerates-more-civilian-deaths-n733336

Dworkin, Anthony. 2013. 'Drones and Targeted Killing: Defining a European Position.' European Council on Foreign Relations. July. Accessed 10 July 2019. https://www.ecfr.eu/page/-/ECFR84_DRONES_BRIEF.pdf

Gallagher, Ryan. 2016. 'Inside Menwith Hill: The NSA's British Base at the Heart of US Targeted Killing.' *The Intercept.* 6 September. Accessed 10 July 2019. https://theintercept.com/2016/09/06/nsa-menwith-hill-targeted-killing-surveillance/

GCHQ. 2011. 'Apparition/Ghosthunter Tasking Info.' *The Intercept.* July. Accessed 10 July 2019. https://assets.documentcloud.org/documents/3089491/Apparition-Ghosthunter-tasking-info.pdf

Grothoff, Christian, and J.M. Porup. 2016. 'The NSA's SKYNET Program May Be Killing Thousands of Innocent People.' *ArsTechnica.* 16 February. Accessed 10 July 2019. https://arstechnica.com/information-technology/2016/02/the-nsas-skynet-program-may-be-killing-thousands-of-innocent-people/

Henckaerts, Jean-Marie, and Louise Doswald-Beck. 2009. *International Committee of the Red Cross, Customary International Humanitarian Law: Volume 1, Rules.* Cambridge: Cambridge University Press.

Isikoff, Michael. 2014. 'After An Errant Drone Strike in Yemen, a Trail of Secret Meetings and US Cash.'. *Yahoo News.* 11 November. Accessed 1 July 2019 https://news.yahoo.com/an-errant-drone-strike-in-yemen--a-secret-white-house-meeting-and-a-bag-of-cash-213034643.html

Jaber, Faisal bin ali. 2016. 'Drone Death Numbers: Has My Family Been Counted?' *The Hill.* 5 July. https://thehill.com/blogs/congress-blog/foreign-policy/286469-drone-death-numbers-has-my-family-been-counted

Jaber, Faisal bin ali. 2013. 'Interview by Reprieve.' *Witness Statement.* Reprieve. 5 November.

Miller, Greg. 2012. 'White House Approves Broader Yemen Drone Campaign.' *Washington Post.* 25 April. Accessed 15 July 2019. https://www.washingtonpost.com/world/national-security/white-house-approves-broader-yemen-drone-campaign/2012/04/25/gIQA82U6hT_story.html?utm_term=.eddc2ecbe32fandwpisrc=nl_headlines

National Security Agency. 2019. 'SKYNET: Applying Advanced Cloud-based Behavior Analytics.' *The Intercept.* 8 May. Accessed 7 July 2019. https://firstlook.org/theintercept/document/2015/05/08/skynet-applying-advanced-cloud-based-behavior-analytics/

Editorial Board. 2013. 'The End of the Perpetual War.' *New York Times*. 23 May. https://www.nytimes.com/2013/05/24/opinion/obama-vows-to-end-of-the-perpetual-war.html

Obama, Barak. 2013. 'Remarks by the President at the National Defense University.' The White House Office of the Press Secretary. 23 May. Accessed 20 June 2019. https://obamawhitehouse.archives.gov/the-press-office/2013/05/23/remarks-president-national-defense-university

Purkiss, Jessica, and Jack Serle. 2017. 'Obama's Covert Drone War in Numbers: Ten Times More Strikes Than Bush.' *The Bureau of Investigative Journalism.* 17 January. Accessed 20 June 2019. https://www.thebureauinvestigates.com/stories/2017-01-17/obamas-covert-drone-war-in-numbers-ten-times-more-strikes-than-bush

Rogers, Christopher. 2016. 'How Should International Law Deal with Doubt in the Era of Drones and Big Data.' *Just Security.* 22 February. Accessed 10 July 2019 https://www.justsecurity.org/29436/ihl-deal-doubt-era-drones-big-data/

Scahill, Jeremy, and Glenn Greenwald. 2019. 'The NSA's Secret Role in the US's Assassination Program.' *The Intercept.* 10 February. Accessed 10 July 2019. https://theintercept.com/2014/02/10/the-nsas-secret-role/

Shane, Scott. 2013. 'Questions on Drone Strike Find Only Silence.' *New York Times*. 22 November: A1.

———. 2016. 'Drone Strike Statistics Answer Few Questions, and Raise Many.' *New York Times*. 3 July.

The Bureau of Investigative Journalism. 2016. 'Naming the Dead.' Accessed 10 June 2019. https://v1.thebureauinvestigates.com/namingthedead/?lang=en

Zenko, Micah. 2016. 'Do Not Believe the US Government's Official Numbers on Drone Strike Civilian Casualties.' *Foreign Policy*. 5 July.

12

Human Judgment in Remote Warfare

JOSEPH CHAPA

Remote warfare describes 'intervention that takes place behind the scenes or at a distance rather than on a traditional battlefield' (Knowles and Watson 2017). But in remote warfare operations, who or what remains at a distance? The impetus for policymakers to pursue policy objectives abroad at low cost and low risk is not a new phenomenon. But the means by which policymakers seek to achieve them does change with technological developments. In the twenty-first century, one of the most noticeable developments in these means has been the advent of remotely piloted aircraft – some call them drones.' This development is especially significant because in previous generations, policymakers established mission objectives from home while their agents – diplomats, soldiers, intelligence officers and others – went out into the operational environment to attempt to achieve those objectives. The advent of remotely piloted aircraft has allowed – in at least some cases – both the policymakers and most of their agents to remain at home while attempting to achieve mission objectives abroad. This supposed removal of the warfighter from the battlespace has raised important ethical questions that have, in turn, spawned a mountain of literature (e.g., Killmister 2008; Strawser 2010; Royakkers and van Est 2010; Galliott 2012; Gregory 2012; Chamayou 2013; Enemark 2014; Kaag and Kreps 2014; Rae and Crist 2014; Gusterson 2015; Himes 2016).

An understudied element of the literature is the role of human judgment in remote warfare. To address this gap, this chapter looks at the relationship between remotely piloted aircraft and human judgment, specifically as it pertains to targeting decisions. The chapter argues that, despite the great physical distances between aircrews and targets, this relatively new technology nevertheless enables crews to apply human judgment in the battlespace as if they were much closer to their weapons' effects.

The ethics of remotely piloted aircraft

Much of the literature on the ethics of remotely piloted aircraft has focused on concerns at the strategic, or policy-level. There are at least two concerns in this category that continue to arise. First, many have argued that voters in liberal democracies are likely to reject military action that results in casualties to their own forces. If remote weapons provide policymakers with military options that will not likely result in casualties to their own forces, then policymakers might have strong political reasons to resort to military force by remote means – perhaps even in cases in which they have strong moral reasons not to. This is often referred to as the 'moral hazard' argument. It suggests that political leaders are perversely incentivised to commit unethical or illegal actions when those actions generate little domestic political risk. Though this argument appears throughout the literature on the ethics of remote weapons, its strongest formulation is in John Kaag's and Sarah Kreps' *Drone Warfare* (see Kaag and Kreps 2012, 2014, 107; Galliott 2012, Chamayou 2013, 189; Brooks 2016, 111).

Another common concern at the strategic level is that remote warfare has enabled powerful states such as the US to employ military force outside areas of active hostilities with relatively little political resistance either domestically or internationally. One possible result is that while al-Qaeda fighters in Afghanistan and Islamic State (ISIS) fighters in Iraq and Syria are lawful combatants, it is not clear whether members of terrorist organisations outside areas of active hostilities (e.g., in Yemen, Somalia, Libya, etc.) are lawful combatants. Though this discussion is about combatant status and not about remote weapons *per se*, it is closely related to the above concern. The ethical concern is that by reducing risk to crews, and therefore reducing political risk to policymakers, policymakers might be incentivised to resort to the unethical use of military force outside areas of active hostilities (see Chamayou 2013, 58; Kaag and Kreps 2014, 2; Enemark 2014, 19-37; Gusterson 2015, 15–21).

These two categories of argument are grounded in the reduced risk to remotely piloted aircraft crews and this reduction in risk is grounded in the physical distance between the crew and their weapons' effects. If the pilot is seven thousand miles from the enemy, she is at no risk of being killed. Because she is at no risk of being killed, policymakers do not face the normal domestic political barriers to the use of military force. Finally, because these strikes are possible without deploying a large force into the country in question, states who employ these systems can potentially conduct violent military actions in a given state without entering into a large-scale war with that state. Much of the literature mentioned above, therefore, is ultimately grounded in the physical distance between crews and targets.

A secondary focus has arisen more recently in a body of literature that distinguishes between physical distance and psychological distance (Asaro 2009; Fitzsimmons and Sangha 2013; Sparrow 2013; Wagner 2014, 1410; Heyns 2016, 11; Lee 2018a). Psychologists as well as ethicists have become increasingly aware that psychological distance is conceptually distinct from physical distance and the two can come apart. Though at great physical distance from their weapons' effects, Predator and Reaper crews, for example, can experience psychological effects as if they were much closer (see Chappelle, Goodman, et al. 2019; Chappelle, McDonald, et al. 2012; Fitzsimmons and Sangha 2013; Maguen, Metzler, et al. 2009). As US Air Force Colonel Joseph Campo (2015) has put it, 'the biggest issue society failed to comprehend was the ability for technology to both separate and connect the warrior to the fight.' In Peter Lee's (2018a) analysis of his interviews with British Royal Air Force Reaper crews, he similarly points to what he calls the 'distance paradox.' Though RAF Reaper crews are physically further from their targets than at any time in the RAF's 100-year history, they are nevertheless emotionally quite close. In his own words, 'aircraft crews had never been so geographically far away from their targets, yet they witnessed and experienced events on the ground in great detail' (Lee 2018a, 113).

Remotely piloted aircraft, however, also raise questions about a third and hitherto under researched sense of distance in war. It might be possible that remotely piloted aircraft crews are able to apply human judgment in the battlespace as if they were quite close, despite the great physical distance between crews and their weapons' effects.

The ethics literature's two-fold focus on physical distance and psychological distance obscures questions about where remote warfare operators can apply human judgment in the battlespace. Psychological distance is a useful conception, but it is limited in that it refers only to the effect violent actions have on the aircraft crews. What I have in mind here compliments, but is crucially distinct from, that conception. Just as the war might affect crews in intimate ways despite the great physical distances involved, those who employ remote weapons might apply human judgment from a relatively close epistemic position despite the great distances involved. In other words, if psychological distance is about the effect the war might have on the crews, the conception of human judgment I have in mind here refers to the effect the crews might have on the war.

One US Air Force Reaper pilot, Lt Clifton, put the relationship between remote crews and their ability to impose human judgment this way.

> [It's] a huge bonus to having that over-the-horizon look – being
> in the [ground control station] vs. being actually in an airplane
> [in] the skies. It's a lot easier to stay calm and stay focused on
> an actual big picture concept instead of just tunnelling in on
> what you see out the window of a fighter jet or what you see in
> the pod of a fighter jet. By physically not being in that
> environment, it keeps the communication between the pilot,
> sensor [operator], and the intel [analysts] a lot smoother, a lot
> more direct, and a lot less hectic to make good decisions and I
> think that's a huge benefit to actually being in [remotely piloted
> aircraft] than being in a manned asset. (Clifton 2019)

Before going further, it is important to bound the scope of this chapter. Those who study 'drones' have sought to keep up with rapid development and proliferation. For instance, a 2017 Center for New American Security study reports that more than 30 countries either have or are developing 'armed drones' (Ewers, Fish, et al. 2017). Likewise, a 2019 New America study finds that 36 countries have 'armed drones' (Bergen, Sterman, et al. 2019). The claims that I make in this paper are not equally applicable across all of these instances for two reasons. This is firstly because the ability for the pilot or crew to impose human judgment depends upon a number of factors about the weapons system in question. ISIS, for example, has employed low-cost quadcopters with 40mm grenades attached after purchase (Gillis 2017; Rassler 2018; Clover and Feng 2017). Suppose a Western military organisation employed such a weapon for local base defence. Such a system does properly fall into the category of 'armed drones,' but it is not at all clear that such a system would provide the operator with sufficient situational awareness to adequately employ human judgment in response to battlefield dynamics.

The second reason is that, because I am here concerned with the relationship between physical distance and human judgment, many of the claims I make will apply directly to systems that a military organisation employs abroad from within its own territory. As Ulrike Franke reports, as of 2017, only a few states – the US, the UK, and China – conduct armed remotely piloted aircraft operations in this way (Franke 2018, 29). At the moment, therefore, my arguments apply most directly to the US and the UK because China's remotely piloted aircraft program is more opaque (see Kania 2018). Moreover, the first-hand narrative accounts I have collected to which I refer below came from US Air Force MQ-9 Reaper crew members and support personnel.[1]

---

[1]   In March of 2019, I interviewed 31 MQ-9 Reaper aircrew members and support personnel at Creech and Shaw Air Force Bases. The interviews were anonymous at the interviewees' request and were intended to provide first-hand perspectives rather than

The conclusions in this paper, however, will likely become more widely applicable as more states begin to operate remotely piloted aircraft from within their own territories.

There is one additional terminological point. 'Physical distance' and 'psychological distance' are less cumbersome than 'distance as it pertains to human judgment' in large part because 'physical' and 'psychological' are such simple and widely understood adjectives. The word 'judgment' does not offer a ready adjective. I propose the more manageable term 'phronetic distance,' which harkens to Aristotle's term 'phronesis,' often translated 'practical wisdom' or 'prudence' (Aristotle and Crisp 2000, 107; Aristotle and Irwin 2000, 345). 'Phronesis' is, for Aristotle, neither knowledge of how to perform a specific task nor is it scientific knowledge. It is a virtue of thought that relies upon reason and enables the one who possesses it to determine what is best for a human being in a wide array of circumstances (Aristotle and Irwin 2000, 8993). Courage is the character trait that enables a virtuous person to act courageously. Temperance is the character trait that enables a virtuous person to act temperately. Phronesis is the trait that enables a virtuous person to know what to do under the circumstances. By 'phronetic distance,' I mean the relative distance between the battlefield and the point of application of human judgment. As I argue below, phronetic distance and physical distance ought to remain conceptually distinct. Though remotely piloted aircraft crews might physically be several thousand miles from the battlefield, their phronetic position is often much closer.

The bin Laden case

Understanding human judgment and distance in remotely piloted aircraft operations is difficult because physical and phronetic distance come apart. In many cases of military technological developments, increases in physical distance between warfighter and weapons' effects correlate with an increase in phronetic distance. In the oft-cited example of early remote weapons, King Henry V's longbowmen at Agincourt are able to engage French knights at a distance. This comes at a marginal increase in phronetic distance. During the fleeting seconds that the weapon is in the air, the longbowmen do not maintain control over it – they have no means of imposing judgment upon where it will impact.

Many military technological developments since Agincourt have followed this model: increases in physical distance result in increases in phronetic distance. Unlike many previous technological developments in which

---

to draw qualitative or quantitative conclusions. The result was more than eight hours of recorded audio and shorthand notes.

increases in physical distance entailed increases in phronetic distance, remotely piloted aircraft have resulted in tremendous increases in physical distance but in relative decreases in phronetic distance.

To see that this is so, consider two cases of modern remote warfare independent of remotely piloted aircraft – namely, the two US attempts on Osama bin Laden's life. In these two cases, the physical distance between the warfighter and the target correlates with phronetic distance.

In 1998, US President Clinton authorised a cruise missile strike against Osama bin Laden following al-Qaeda 's bombings of US embassies in Kenya and Tanzania. The US Navy prosecuted the attack against what the US believed to be bin Laden's location near Khost, Afghanistan with ship-fired Tomahawk cruise missiles from the Arabian Sea (Kean, Hamilton, et al. 2004, 116–117). Bin Laden had indeed planned on going to Khost where he likely would have been killed in the strike. But, as Lawrence Wright (2011, 321–322) describes, on the way there, in a car with his friends, bin Laden said:

> 'Where do you think, my friends, we should go ... Khost or Kabul?'

> His bodyguard and others voted for Kabul, where they could visit friends.

> 'Then, with God's help, let us go to Kabul,' bin Laden decreed – a decision that may have saved his life.

In this case, the naval surface warfare officer in the Arabian Sea responsible for launching the cruise missile was some five hundred miles from the target area. This distance, though considerably closer than the remotely piloted aircraft pilot thousands of miles away, is still applying military force while remaining outside the theatre of operations. But, crucially in the bin Laden case, the surface warfare officer has no means of imposing his or her judgment after the missile is launched. Just as King Henry's longbowmen accepted an increase in phronetic distance, the cruise missile also imposes an increase in phronetic distance. For the longbowmen, this increase was marginal – the arrow's flight time is a matter of single digit seconds. Like the longbowmen's arrows, the cruise missile can neither be recalled nor can they be redirected once launched and its flight time is four to six hours long (Navy 2018; Shane 2016). And, of course, even if the missiles could have been redirected, the surface warfare officer has no intelligence feedback loop to alert him to the fact that the intelligence reporting was mistaken. Though the physical distance was significant, the application of human judgment in response to real-time dynamics on the ground is completely absent. In this case, the increase in physical distance entails an increase in phronetic distance.

Compare this 1998 event against the US's 2011 raid that killed Osama bin Laden in Abbottabad, Pakistan. US President Obama opted for a 'capture or kill' mission conducted by special operations forces that ultimately led to bin Laden's death. What is crucial for the present discussion is that the forces in the helicopters and on the ground had both the capability and the authority to employ their judgment in response to real-time dynamics. The raid provides two important examples. The first helicopter to arrive in the compound had planned to hover while the operators inside fast roped into the compound. But the solid walls of the compound affected the airflow differently from the chain-link fence within which the team had practiced. In response to the unexpected and debilitating air currents, the helicopter pilot had to put the helicopter down in the compound, ultimately in a forced landing that severely damaged the aircraft. The pilot of the second helicopter saw the first helicopter's landing and was unsure whether the landing and damage were the result of enemy fire or mechanical problems. The second pilot, therefore, decided to land outside the compound forcing the SEALs to run into the compound from there – both were major deviations from the original plan (Schmidle 2011; Swinford 2011). In this instance, the team relied, not upon scripted orders from higher headquarters, nor on communications reach back. They employed human judgment in response to real-time battlefield dynamics.

Second, and more importantly, the room from which US Cabinet and other officials – including President Obama – watched the raid lost communications with the raid force for some 20–25 minutes (Swinford 2011) over half the time the team was on the ground (Schmidle 2011). During this crucial period of the 'kill or capture' mission, the raid team chose, based on real-time dynamics on the ground, to kill rather than to capture bin Laden. Again, they relied upon human judgment. Leon Panetta, then Director of the CIA, told reporters that 'It was a firefight going up that compound. And by the time they got to the third floor and found bin Laden, I think it – this was all split-second action on the part of the SEALs' (Swinford 2011).[2] Here, the fact that the special operators are in close physical proximity to the battlefield and to their target enables them to apply human judgment from a relatively close epistemic position. Their decreased physical distance to the target entails a decrease in phronetic distance.

---

[2]   This is a contested point. In Schmidle's account, he cites a special operations officer who claims that 'There was never any question of detaining or capturing him—it wasn't a split-second decision. No one wanted detainees.' Because I am after the conceptual distinction between physical and phronetic distance, this disagreement can be set to one side.

In each of these two cases, physical distance correlates with phronetic distance. The naval surface warfare officer responsible for the 1998 cruise missiles is physically 500 miles away from his intended target, and the point of application of human judgment is at his physical fingertips. His ability to react to apply human judgment in response to real-time dynamics is constrained by the technological limitations of the weapon and by the officer's physical dislocation from the target area. The special operators in the Abbottabad raid, however, are able to perceive real-time battlefield dynamics and apply human judgment in response because, among other things, they are physically present in the target area.

The task in the remainder of this chapter is to show that unlike in these two examples, in remotely piloted aircraft operations, increases in physical distance do not necessarily correlate with increases in phronetic distance.

Phronetic distance and remotely piloted aircraft

At first glance, it might look as though the phronetic distance from which remotely piloted aircraft crews apply human judgment is similar to phronetic distance in the standoff cruise missile case. Our intuitions in response to this question have unfortunately been primed by widespread misconceptions in both the popular and scholarly literature on 'drones.' We are often told that these systems are robotic (Schneider and Macdonald 2017; Coeckelbergh 2013, 90; Royakkers and van Est 2010, 289; Sharkey 2013, 797); and that they fall into the class of autonomous or semi-autonomous weapons (Kaag and Kreps 2014, vii; Brunstetter and Braun 2011, 338). These descriptors, 'robotic' and 'semi-autonomous,' are more apt for the cruise missile. It flies a pre-planned route toward a pre-designated target and the human operator cannot intervene post-launch. Neither of these claims obtain for the remotely piloted aircraft.

Unfortunately, published first-hand accounts from remotely piloted aircraft crews that might either confirm or rebut these claims are few. There are just two US pilot memoirs of which I am aware and a third written by a US intelligence analyst (Martin and Sasser 2010; McCurley 2017; Velicovich and Stewart 2017).[3] Peter Lee has also helpfully collected first-hand accounts from British Royal Air Force Reaper crews in his 2018 book, *Reaper Force* (Lee 2018b). Campo's study provides an important, if often overlooked, insight here. Though the primary focus of his study was the psychological

---

[3]   Martin's memoir is particularly contentious within the US Air Force Reaper (and formerly Predator) community. See, for example, Byrnes, C. M. W. 2018. Review: 'We Kill Because We Can: From Soldiering to Assassination in the Drone Age.' *Air and Space Power Journal,* 32.

effects on remote warfare crews, he did ask US Air Force Predator and Reaper about instances in which they had intervened to stop or delay a strike. Among his more than one hundred interviewees, twenty-two subjects provided narrative accounts in which they applied human judgment to intervene to stop or delay a strike. In Campo's words,

> All twenty-two stories were remarkably similar. In each story, the aircrew were directed to strike a target, but something just 'did not feel right' to them regarding the situation, the target identification, or the surrounding area. In every case, the aircrew took positive steps to understand the situation, develop their own mental model of the battlespace, and then recommend (or demand) a different course of action besides immediate weapons engagement via [remotely piloted aircraft]. All twenty-two individuals steadfastly believe that had they simply followed directions without delay or critical inquiry, collateral damage or civilian casualties were nearly assured. (Campo 2015, 7–8)

Though Campo does not use the words 'human judgment,' his description is relevantly similar to my description of human judgment above. The theme Campo observed reappeared anecdotally in my own discussions with Predator and Reaper crews. One US Air Force pilot, Captain Andy, told me about a case in which the Airman attached to the ground team who was directing the strike – a joint terminal attack controller, or JTAC (pronounced 'jay-tack') – was confused and disoriented while taking enemy fire:

> The friendlies were getting shot at. Both sides were, I think, 75 meters apart. We got a 9 line [attack briefing from the JTAC] to shoot friendly forces. The sensor [operator] was like, "holy crap. This is just not right." The hairs on the back of the neck stood up, then we correlated more, and then we told the JTAC, "hey, you gave us a 9-line for yourself," […] "the grids are over here." You're not going to get that with a robot. […] You'll give [the robot] a grid and tell them to shoot it and [it's] going to shoot it."

This account, and others like it, run counter to the received wisdom on how remote warfighters will respond to battlefield dynamics. For example, in his 2013 chapter, 'War Without Virtue,' Rob Sparrow (2013, 100–101) anticipates that 'since the [remote] operators are not in any danger, it is more plausible to expect them to follow orders from other people who may be geographically distant and also to wait for orders to follow.'

Lt Clifton, a Reaper pilot and formerly a sensor operator, disagrees. He mentions three times that he 'pushed back' against the JTAC's instructions.

> Those three strikes would have been legal based on actions, locations, and what was observed, but because of other factors which I voiced up (I wasn't comfortable with the shot) [...] You just don't have a warm fuzzy because you don't have all the details necessary. [...] I've had three specific occasions where I voiced it up and the JTAC said, "copy that, we'll hold off". (Clifton 2019)

Lt Clifton went on to say that 'it's a two-way process between JTACs and aircrew. JTACs can tell us "cleared hot" all day long, and give us orders to strike, but of course as aircrew we don't have to because the weapon is ultimately our responsibility' (Clifton 2019).

An instructor sensor operator, Technical Sergeant Megan, put it this way:

> There [have] been several situations where I would say the conversation between the pilot in command or the crew and the JTAC [...] is – I don't want to say "heated," but they feel like this is what needs to be done and the crew [says], 'we're not comfortable with that' for whatever reason. ... At the end of the day, this is [the pilot's] weapon. This is our aircraft. This is what we're comfortable with doing and this is what we're not comfortable with doing. [...] At the end of the day, I'd say most of our crews are very good at standing up for that. (Megan 2019)

I asked another instructor sensor operator named Master Sergeant Sean if he had ever experienced a moral dilemma in the seat. He said:

> I wouldn't say that I've ever had a moral dilemma [...] Just because typically we work so well as a crew just between myself as the sensor operator and the pilot, that we're able to come to a reasonable solution [...] JTACs are pretty receptive when we push back on them and say, "hey, we're just not comfortable with the strike. Can we just, you know, hold off a little bit?"

He went on to say:

> I've had several [instances] where we weren't comfortable with a certain strike just because we were worried about CIVCAS [civilian casualties] and things like that so we pushed back to the JTAC and ended up waiting, and lo and behold, we were able to eliminate the target in clear terrain with no CIVCAS. (Sean 2019)

Though the resounding claims from the US Reaper crewmembers interviewed suggest that they do have the capability to apply human judgment, there are still constraints on the crews' ability to impose human judgment.

Phronetic distance in traditionally piloted aircraft

The above quotations do not suggest that the remote crews can impose human judgment to the same degree that the special operators did in the bin Laden raid. One of the most significant differences between the two is the difference between their epistemic positions. To see a target through a targeting pod at 20,000 feet does provide the aircrew with greater awareness than was available in the 1998 cruise missile case. But the remotely piloted aircraft crew's epistemic state is still far different from that of the soldier on the ground. Retired US Army General Stanley McChrystal, former commander of coalition forces in Afghanistan, put the epistemic concern this way: 'Because if you see things in 2D, a photograph or a flat screen, you think you know what's going on, but you don't know what's going on, you only know what you see in two dimensions' (quoted in Kennebeck 2017). So how are we to understand phronetic distance in remotely piloted aircraft? If the phronetic distance that is relevant in remotely piloted aircraft operations is neither like previous generates of long-distance weapons nor like traditional warfighters on the ground, perhaps the more apt point of comparison is traditionally piloted aircraft. That is, though this relatively recent technological development has had profound impacts on physical distance and psychological distance, perhaps phronetic distance in air operations is more continuous.

I spoke with Captain Shaun and Technical Sergeant Megan in a ground control station while they flew an operational mission over Afghanistan. Captain Shaun has experience both as a Reaper pilot and as a MC-12 Liberty pilot – an unarmed, traditionally piloted, propeller driven airplane used for intelligence, surveillance and reconnaissance. While flying the MC-12 in Afghanistan, the numerous intelligence analysts and ground personnel watching his video feed reported two people emplacing an improvised explosive device (IED) in a culvert under a road. The various participants in the operation started preparing an attack briefing for another aircraft. Captain

Shaun and his crew were not convinced that what they saw was an IED emplacement and repeatedly intervened in the momentum that was building toward a strike. In Captain Shaun's words, 'it didn't feel right. We stalled the kill chain multiple times.' The 'kill chain' is the US military's shorthand for the dynamic targeting process, consisting the six steps, 'find, fix, track, target, engage, and assess' (USAF 2019) Eventually, Captain Shaun said:

> The two people we were watching ended up walking up to two full-grown adults. Once we saw the relative size, we knew the two people we had been watching were kids. They [had been] pulling sticks out of a culvert to get the water to flow. If we hadn't stalled the kill chain, who knows what would have happened? (Shaun 2019)

In my view, this is undoubtedly a case in which the crew applied human judgment in the battlespace. In this case, the phronetic distance correlates with physical distance. Captain Shaun's physical and phronetic position is 15,000 feet above the target and he is capable of observing and intervening from that position. Had he been a soldier on the ground, his epistemic position would have been different, the fact that the two people were children would have been more obvious, and his ability to apply human judgment strengthened.

When I asked Captain Shaun about the differences between his ability to apply human judgment in the traditionally piloted MC-12 and in the remotely piloted Reaper, he said interrupting the kill chain is even easier in the Reaper because he is now responsible, not just for the camera providing the situational awareness, but also for the weapon. 'I can say 'I'm the A-code [the pilot in command]. It's my weapon. My sensor operator doesn't like it. We're not doing it.' Sgt Megan added, 'You have to have that level of respect that it's a human life you're taking. I'll still do it for the right reasons, but it has to be for the right reasons.' But as we have already seen, there are some conditions under which one's position half a world away might improve one's epistemic position, perhaps especially when friendly forces are taking fire.

Conclusion: Empowering judgment

If the first limitation on the remotely piloted aircraft crew's application of human judgment in the battlespace is their epistemic position, the second is the organisational constraints on their autonomy. This is a question, not of technological capability, but of organisational culture, doctrine, and training. The technological capability – the visualisation of the battlespace via high-resolution cameras in multiple segments of light spectrum; the long loiter times over the target area; and the integrated network of operators, intelligence analysts, and commanders – is a necessary, but insufficient condition for applying human judgment in the battlespace.

For the last few decades, many Western militaries, including NATO on the whole, have moved toward a concept of 'mission command' according to which commanders issue mission-type orders with an emphasis on the commander's intent to 'thereby empowering agile and adaptive [subordinate] leaders with freedom to conduct operations' (Roby and Alberts 2010, xvi; Scaparrotti and Mercier 2018, 2017, 6, 18, 37; Storr 2003). The freedom to conduct operations that is so central to mission command consists in the freedom to employ human judgment in the battlespace. In this approach, subordinate commanders, to include pilots in command, will retain the authority required to apply human judgment even in complex and difficult circumstances. A recurring, though not universal, theme in my interviews with Reaper crews was that commanders at the squadron level and above would support pilots' decisions when those pilots employed human judgment – and especially restraint – in the battlespace. Though the interviewees were with American Reaper crewmembers, it is noteworthy that Reaper crewmembers from the UK, France, Italy, Australia, and The Netherlands train alongside one another – perhaps inculcating this empowered approach to human judgment (Tran 2015; Murray 2013; Stevenson 2015; Fiorenza 2019). As these systems continue to proliferate, however, it is not yet clear whether all the states that will operate them will continue to value aircrew autonomy.

Finally, as military technology continues to develop it will be important to compare the application of human judgment in remote weapons employment to potential future use of autonomous weapons. In many instances, it has been human judgment, rather than targeting systems, that have identified errors and prevented catastrophic strikes. As militaries continue to develop artificial intelligence systems and apply them in the targeting process, they risk eroding the crucial application of human judgment in some situations. If nothing else, this discussion of human judgment in the battlespace should motivate developers and military commanders, not merely to ask which military tasks can be automated, but also to ask where in the battlespace human judgment ought to be preserved.

The views expressed in this chapter are those of the author and do not necessarily reflect those of the US Air Force, the Department of Defense, or the US Government.

References

Alberts, David S., Reiner K. Huber and James Moffat. 2010. *NATO NEC C2 Maturity Model. SAS-065.* Washington: DoD Command and Control Research Program.

Aristotle and Crisp, Roger. 2000. *Nicomachean ethics.* Cambridge, U.K., Cambridge University Press.

Aristotle and Irwin, Terence. 2000. *Nicomachean Ethics.* Indianapolis, Hackett Publishing Company Inc.

Asaro, Peter. 2009. Modeling the moral user. *IEEE Technology and Society Magazine,* 28: 20–24.

Bergen, Peter, David Sternman, Alyssa Sims, Albert Ford, and Christopher Mellon. 2016 (updated 2019). 'World of Drones: Examining the Proliferation, Development, and Use of Armed Drones.' New America.

Brooks, Rosa. 2016. *How everything became war and the military became everything: tales from the Pentagon.* New York: Simon and Schuster.

Brunstetter, Daniel, and Megan Braun. 2011. 'The Implications of Drones on the Just War Tradition.' *Ethics and International Affairs,* 25: 337–358.

Byrnes, Captain Michael W. 2018. 'Review: We Kill Because We Can: From Soldiering to Assassination in the Drone Age.' *Air and Space Power Journal,* 32.

Campo, Joseph L., 2015. 'Distance in War: The Experience of MQ-1 and MQ-9 Aircrew.' *Air and Space Power Journal.*

Catalano Ewers, Elisa, Lauren Fish, Michael C. Horowitz, Alexandra Sander and Paul Scharre. 2017. 'Drone Proliferation: Policy Choices for the Trump Administration.' *Papers for The President.* Washington: Center for New American Security.

Chamayou, Gregoire. 2013. *A Theory of The Drone.* New York, The New Press.

Chappelle, Wayne, Tanya Goodman, Laura Reardon and Lillian Prince. 2019. 'Combat and operational risk factors for post-traumatic stress disorder symptom criteria among United States air force remotely piloted aircraft "Drone" warfighters.' *Journal of Anxiety Disorders,* 62**:** 86–93.

Chappelle, Wayne, Kent McDonald, Billy Thompson, and Julie Swearengen. 2012. *Prevalence of High Emotional Distress and Symptoms of Post-Traumatic Stress Disorder in US Air Force Active Duty Remotely Piloted Aircraft Operators: 2010 USAFAM Survey Results. Final Technical Report.* Wright Patterson Air Force Base, OH: Air Force Research Laboratory.

Clifton, Lt. 'Interview with MQ-9 Reaper Personnel.' By Chapa, Joseph. 14 March 2019.

Clover, Charles, and Emily Feng. 2017. 'Isis use of hobby drones as weapons tests Chinese makers.' *Financial Times*. 11 December.

Coeckelbergh, Mark. 2013. 'Drones, information technology, and distance: mapping the moral epistemology of remote fighting.' *Ethics and Information Technology,* 15**:** 87– 98.

Enemark, Christian. 2014. *Armed drones and the ethics of war: military virtue in a post-heroic age.* London: Routledge.

Fiorenza, Nicholas. 2019. 'RNLAF Reaper operators train in US.'*Jane's Defence Weekly*, 21 January.

Fitzsimmons, Scott, and Karina Sangha. 2013. 'Killing in High Definition: Combat Stress among Operators of Remotely Piloted Aircraft.' *Technology*, 12: 289– 292.

Franke, Ulrike. 2018. *The unmanned revolution: how drones are revolutionising warfare.* ProQuest Dissertations Publishing.

Galliott, Jai C., 2012. 'Uninhabited Aerial Vehicles and The Asymmetry Objection: A Response to Strawser.' *Journal of Military Ethics,* 11(1)**:** 58–66.

Gillis, Jonathan. 2017. 'In over their heads: US ground forces are dangerously unprepared for enemy drones.' *War on The Rocks*. 30 May.

Gregory, Derek. 2012. 'From a View to a Kill.' *Theory, Culture and Society,* 28(7– 8)**:** 188–215.

Gusterson, Hugh. 2015. *Drone: Remote Control Warfare.* Cambridge, MA: MIT Press.

Heyns, Christof. 2016. 'Human rights and the use of autonomous weapons systems (AWS) during domestic law enforcement.' *Human Rights Quarterly,* 38: 350–378.

Himes, Kenneth R., 2016. *Drones and the Ethics of Targeted Killing.* Lanham, Rowman and Littlefield.

HM Government. 2017. 'Future of Command and Control.' The Development, Concepts and Doctrine Centre.

Kaag, John, and Sarah Kreps. 2012. 'The Moral Hazard of Drones.' *The New York Times,* 22 July.

————. 2014. *Drone Warfare.* Cambridge: Polity.

Kania, Elsa. 2018. 'The PLA's Unmanned Aerial Systems: New Capabilities for a "New Era" of Chinese Military Power.' In *China Aerospace Studies Institute,* edited by Brendan Mulvaney. Montgomery: Air University.

Kean, Thomas H,, and Lee Hamilton, Richard Ben-Veniste, Bob Kerrey, Fred F. Fielding, John F. Lehman, Jamie F. Gorelick, Timothy J. Roemer, Slade Gorton and James R. Thompson. 2004. *The 9/11 Commission Report.* National Commission on Terrorist Attacks upon the United States.

Kennebeck, Sonia. 2016. *National Bird. Washington, DC: Ten Forward Films.*

Killmister, Suzy. 2008. Remote Weaponry: The Ethical Implications. *Journal of Applied Philosophy,* 25(2): 121–133.

Knowles, Emily, and Abigail Watson, A. 2017. *All Quiet on the ISIS Front: British Secret Warfare in an Information Age.* Remote Control Project, Oxford Research Group.

Lee, Peter. 2018a. 'The distance paradox: reaper, the human dimension of remote warfare, and future challenges for the RAF.' *Air Power Review,* 21(3): 106–130.

Lee, Peter. 2018b. *Reaper Force: The Inside Story of Britain's Drone Wars.* London: John Blake Publishing.

Maguen, Shira, Thomas J. Metzler, Brett T. Litz, Karen H. Seal, Sara J. Knight, and Charles R. Marmar. 2009. 'The impact of killing in war on mental health symptoms and related functioning.' *Journal of Traumatic Stress,* 22(5): 435–443.

Martin, Matt J., and Charles W. Sasser. 2010. *Predator: The remote-control air war over Iraq and Afghanistan: A pilot's story*. Zenith Press.

McCurley, Mark T., 2017. *Hunter Killer: Inside the Lethal World of Drone Warfare*. Atlantic Books.

Megan, T. 'Interview with MQ-9 Reaper Personnel.' By Chapa, Joseph. 14 March 2019.

Murray, Airman 1st Class Leah. 'Italians learn to fly RPAs at Holloman.' News release. 13 June, 2013, https://www.holloman.af.mil/Article-Display/Article/317370/italians-learn-to-fly-rpas-at-holloman/

Navy. 2018. 'Tomahawk Cruise Missile.' News release. 26 April. https://www.navy.mil/Resources/Fact-Files/Display-FactFiles/Article/2169229/tomohawk-cruise-missile/

Rae, James Deshaw. 2014. *Analyzing the drone debates: targeted killing, remote warfare, and military technology.* Basingstoke: Palgrave Macmillan.

Rassler, Don. 2018. *The Islamic State and Drones: Supply, Scale, and Future Threats.* West Point: Combating Terrorism Center at West Point.

Royakkers, Lambèr, and Rinie Van Est. 2010. 'The cubicle warrior: the marionette of digitalized warfare.' *Ethics and Information Technology,* 12(3): 289–296.

Scaparrotti, Curtis M., and Denis Mercier. 2018. 'Framework for Future Alliance Operations.' *Keeping The Edge.* NATO.

Schmidle, Nicholas. 2011. 'Getting Bin Laden'. *The New Yorker,* 8 August.

Schneider, Jacquelyn, and Julia Macdonald. 2017. 'Why Troops Don't Trust Drones: The 'Warm Fuzzy' Problem.' *Foreign Affairs*. 20 December.

Sean, Major. 2019. 'Interview with MQ-9 Reaper Personnel.' By Chapa, Joseph. 14 March.

Shane, Scott. 2016. *Objective Troy: A Terrorist, A President, and The Rise of The Drone*. Seal Books.

Sharkey, Noel E., 2013. 'The evitability of autonomous robot warfare.' *International Review of the Red Cross,* 94(886): 787–799.

Shaun, Captain. 2019. 'Interview with MQ-9 Reaper Personnel.' By Chapa, Joseph. O. 15 March.

Sparrow, Robert. 2013. 'War Without Virtue?' In Strawser, Bradley. J. (ed.) *Killing by Remote Control.* Oxford: Oxford University Press.

Stevenson, Beth. 2015. 'RAAF begins Reaper training in USA.' *Flight Global,* 23 February.

Storr, Jim. 2003. 'A Command Philosophy for the Information Age: The Continuing Relevance of Mission Command.' *Defence Studies,* 3(3): 119–129.

Strawser, Bradley J., 2010. 'Moral Predators: The Duty to Employ Uninhabited Aerial Vehicles.' *Journal of Military Ethics,* 9(4): 342–368.

Swinford, Steven. 2011. 'Osama bin Laden Dead: Blackout During Raid on bin Laden Compound.' *The Telegraph*. 4 May.

Tran, Pierre. 2015. 'UK, France Discuss Reaper Pilot Training.' *Defense News*. 3 June.

USAF. 2019. Annex 3-60: Targeting. In *Education, C. E. L. C. F. D. D. A.,* edited by Maxwell Air Force Base, AL.

Velicovich, Brett, and Christopher S. Stewart. 2017. *Drone Warrior: An Elite Soldier's Inside Account of the Hunt for America's Most Dangerous Enemies*. Harper Collins.

Wagner, Markus. 2014. 'The Dehumanizatino of International Humanitarian Law: Legal, Ethical, and Political Implications of Autonomous Weapon Systems.' *Vand. J. Transnat'l L.,* 47**:** 1371.

Wright, Lawrence. 2011. The Looming Tower: al-Qaeda's Road to 9/11 London: Penguin

13

The Future of Remote Warfare? Artificial Intelligence, Weapons Systems and Human Control

INGVILD BODE AND HENDRIK HUELSS

The use of force exercised by the militarily most advanced states in the last two decades has been dominated by 'remote warfare', which, at its simplest, is a 'strategy of countering threats at a distance, without the deployment of large military forces' (Oxford Research Group cited in Biegon and Watts 2019, 1). Although remote warfare comprises very different practices, academic research and the broader public pays much attention to drone warfare as a very visible form of this 'new' interventionism. In this regard, research has produced important insights into the various effects of drone warfare in ethical, legal, political, but also social and economic contexts (Cavallaro, Sonnenberg and Knuckey 2012; Sauer and Schörnig 2012; Casey-Maslen 2012; Gregory 2015; Hall and Coyne 2013; Schwarz 2016; Warren and Bode 2015; Gusterson 2016; Restrepo 2019; Walsh and Schulzke 2018). But current technological developments suggest an increasing, game-changing role of artificial intelligence (AI) in weapons systems, represented by the debate on emerging autonomous weapons systems (AWS). This development poses a new set of important questions for international relations, which pertain to the impact that increasingly autonomous features in weapons systems can have on human decision-making in warfare – leading to highly problematic ethical and legal consequences.

In contrast to remote-controlled platforms such as drones, this development refers to weapons systems that are AI-driven in their critical functions. That is weapons that process data from on-board sensors and algorithms to 'select (i.e., search for or detect, identify, track, select) and attack (i.e., use force against, neutralise, damage or destroy) targets without human intervention'

(ICRC 2016). AI-driven features in weapons systems can take many different forms but clearly depart from what might be conventionally understood as 'killer robots' (Sparrow 2007). We argue that including AI in weapons systems is important not because we seek to highlight the looming emergence of fully autonomous machines making life and death decisions without any human intervention, but because human control is increasingly becoming compromised in human-machine interactions. AI-driven autonomy has already become a new reality of warfare. We find it, for example, in aerial combat vehicles such as the British Taranis, in stationary sentries such as the South Korean SGR-A1, in aerial loitering munitions such as the Israeli Harop/Harpy, and in ground vehicles such as the Russian Uran-9 (see Boulanin and Verbruggen 2017). These diverse systems are captured by the (somewhat problematic) catch-all category of autonomous weapons, a term we use as a springboard to draw attention to present forms of human-machine relations and the role of AI in weapons systems short of full autonomy.

The increasing sophistication of weapons systems arguably exacerbates trends of technologically mediated forms of remote warfare that have been around for some decades. The decisive question is how new technological innovations in warfare impact human-machine interactions and increasingly compromise human control. The aim of our contribution is to investigate the significance of AWS in the context of remote warfare by discussing, first, their specific characteristics, particularly with regard to the essential aspect of distance and, second, their implications for 'meaningful human control' (MHC), a concept that has gained increasing importance in the political debate on AWS. We will consider MHC in more detail further below.

We argue that AWS increase fundamental asymmetries in warfare and that they represent an extreme version of remote warfare in realising the potential absence of immediate human decision-making on lethal force. Furthermore, we examine the issue of MHC that has emerged as a core concern for states and other actors seeking to regulate AI-driven weapons systems. Here, we also contextualise the current debate with state practices of remote warfare relating to systems that have already set precedents in terms of ceding meaningful human control. We will argue that these incremental practices are likely to change use of force norms, which we loosely define as standards of appropriate action (see Bode and Huelss 2018). Our argument is therefore less about highlighting the novelty of autonomy, and more about how practices of warfare that compromise human control become accepted.

Autonomous weapons systems and asymmetries in warfare

AWS increase fundamental asymmetries in warfare by creating physical, emotional and cognitive distancing. First, AWS increase asymmetry by creating physical distance in completely shielding their commanders/ operators from physical threats or from being on the receiving end of any defensive attempts. We do not argue that the physical distancing of combatants has started with AI-driven weapons systems. This desire has historically been a common feature of warfare – and every military force has an obligation to protect its forces from harm as much as possible, which some also present as an argument for remotely-controlled weapons (see Strawser 2010). Creating an asymmetrical situation where the enemy combatant is at the risk of injury while your own forces remain safe is, after all, a basic desire and objective of warfare.

But the technological asymmetry associated with AI-driven weapon systems completely disturbs the 'moral symmetry of mortal hazard' (Fleischman 2015, 300) in combat and therefore the internal morality of warfare. In this type of 'riskless warfare, […] the pursuit of asymmetry undermines reciprocity' (Kahn 2002, 2). Following Kahn (2002, 4), the internal morality of warfare largely rests on 'self-defence within conditions of reciprocal imposition of risk.' Combatants are allowed to injure and kill each other 'just as long as they stand in a relationship of mutual risk' (Kahn 2002, 3). If the morality of the battlefield relies on these logics of self-defence, this is deeply challenged by various forms of technologically mediated asymmetrical warfare. It has been voiced as a significant concern in particular since NATO's Kosovo campaign (Der Derian 2009) and has since grown more pronounced through the use of drones and, in particular, AI-driven weapons systems that decrease the influence of humans on the immediate decision-making of using force.

Second, AWS increase asymmetry by creating an emotional distance from the brutal reality of wars for those who are employing them. While the intense surveillance of targets and close-range experience of target engagement through live pictures can create intimacy between operator and target, this experience is different from living through combat. At the same time, the practice of killing from a distance triggers a sense of deep injustice and helplessness among those populations affected by the increasingly autonomous use of force who are 'living under drones' (Cavallaro, Sonnenberg and Knuckey 2012). Scholars have convincingly argued that 'the asymmetrical capacities of Western – and particularly US forces – themselves create the conditions for increasing use of terrorism' (Kahn 2002, 6), thus 'protracting the conflict rather than bringing it to a swifter and less bloody end' (Sauer and Schörnig 2012, 373; see also Kilcullen and McDonald Exum 2009; Oudes and Zwijnenburg 2011).

This distancing from the brutal reality of war makes AWS appealing to casualty-averse, technologically advanced states such as the USA, but potentially alters the nature of warfare. This also connects well with other 'risk transfer paths' (Sauer and Schörnig 2012, 369) associated with practices of remote warfare that may be chosen to avert casualties, such as the use of private military security companies or working via airpower and local allies on the ground (Biegon and Watts 2017). Casualty aversion has been mostly associated with a democratic, largely Western, 'post-heroic' way of war depending on public opinion and the acceptance of using force (Scheipers and Greiner 2014; Kaempf 2018). But reports about the Russian aerial support campaign in Syria, for example, speak of similar tendencies of not seeking to put their own soldiers at risk (The Associated Press 2018). Mandel (2004) has analysed this casualty aversion trend in security strategy as the 'quest for bloodless war' but, at the same time, noted that warfare still and always includes the loss of lives – and that the availability of new and ever more advanced technologies should not cloud thinking about this stark reality.

Some states are acutely aware of this reality as the ongoing debate on the issue of AWS at the UN Convention on Certain Conventional Weapons (UN-CCW) demonstrates. It is worth noting that most countries in favour of banning autonomous weapons are developing countries, which are typically less likely to attend international disarmament talks (Bode 2019). The fact that they are willing to speak out strongly against AWS makes their doing so even more significant. Their history of experiencing interventions and invasions from richer, more powerful countries (such as some of the ones in favour of AWS) also reminds us that they are most at risk from this technology.

Third, AWS increase cognitive distance by compromising the human ability to 'doubt algorithms' (see Amoore 2019) in terms of data outputs at the heart of the targeting process. As humans using AI-driven systems encounter a lack of alternative information allowing them to substantively contest data output, it is increasingly difficult for human operators to doubt what 'black box' machines tell them. Their superior data processing capacity is exactly why target identification via pattern recognition in vast amounts of data is 'delegated' to AI-driven machines, using, for example, machine-learning algorithms at different stages of the targeting process and in surveillance more broadly.

But the more target acquisition and potential attacks are based on AI-driven systems as technology advances, the less we seem to know about how those decisions are made. To identify potential targets, countries such as the USA (e.g. SKYNET programme) already rely on meta-data generated by machine-learning solutions focusing on pattern of life recognition (The Intercept 2015; see also Aradau and Blanke 2018). However, the lacking ability of humans to retrace how algorithms make decisions poses a serious ethical, legal and political problem. The inexplicability of algorithms makes it harder for any human operator, even if provided a 'veto' or the power to intervene 'on the loop' of the weapons system, to question metadata as the basis of targeting and engagement decisions. Notwithstanding these issues, as former Assistant Secretary for Homeland Security Policy Stewart Baker put it, 'metadata absolutely tells you everything about somebody's life. If you have enough metadata, you don't really need content', while General Michael Hayden, former director of the NSA and the CIA emphasises that '[w]e kill people based on metadata' (both quoted in Cole 2014).

The desire to find (quick) technological fixes or solutions for the 'problem of warfare' has long been at the heart of debates on AWS. We have increasingly seen this at the Group of Governmental Experts on Lethal Autonomous Weapons Systems (GGE) meetings at the UN-CCW in Geneva when countries already developing such weapons highlight their supposed benefits. Those in favour of AWS (including the USA, Australia and South Korea) have become more vocal than ever. The USA claimed that such weapons could actually make it easier to follow international humanitarian law by making military action more precise (United States 2018). But this is a purely speculative argument at present, especially in complex, fast-changing contexts such as urban warfare. Key principles of international humanitarian law require deliberate human judgements that machines are incapable of (Asaro 2018; Sharkey 2008). For example, the legal definition of who is a civilian and who is a combatant is not written in a way that could be easily programmed into AI, and machines lack the situational awareness and ability to infer things necessary to make this decision (Sharkey 2010).

Yet, some states seem to pretend that these intricate and complex issues are easily solvable through programming AI-driven weapons systems in just the right way. This feeds the technological 'solutionism' (Morozov 2014) narrative that does not appear to accept that some problems do not have technological solutions because they are inherently political in nature. So, quite apart from whether it is technologically possible, do we want, normatively, to take out deliberate human decision making in this way?

This brings us to our second set of arguments concerned with the fundamental questions that introducing AWS into practices of remote warfare pose to human-machine interaction.

The problem of meaningful human control

AI-driven systems signal the potential absence of immediate human decision-making on lethal force and the increasing loss of so-called meaningful human control (MHC). The concept of MHC has become a central focus of the ongoing transnational debate at the UN-CCW. Originally coined by the non-governmental organisation (NGO) Article 36 (Article 36 2013, 36; see Roff and Moyes 2016), there are different understandings of what meaningful human control implies (Ekelhof 2019). It promises resolving the difficulties encountered when attempting to define precisely what autonomy in weapons systems is but also meets somewhat similar problems in its definition of key concepts. Roff and Moyes (2016, 2–3) suggest several factors that can enhance human control over technology: technology is supposed to be predictable, reliable, transparent; users should have accurate information; there is timely human action and a capacity for timely intervention, as well as human accountability. These factors underline the complex demands that could be important for maintaining MHC but how these factors are linked and what degree of predictability or reliability, for example, are necessary to make human control meaningful remains unclear and these elements are underdefined.

In this regard, many states consider the application of violent force without any human control as unacceptable and morally reprehensible. But there is less agreement about various complex forms of human-machine interaction and at what point(s) human control ceases to be meaningful. Should humans always be involved in authorising actions or is monitoring such actions with the option to veto and abort sufficient? Is meaningful human control realised by engineering weapons systems and AI in certain ways? Or, more fundamentally, is human control that consists of simply executing decisions based on indications from a computer that are not accessible to human reasoning due to the 'black-boxed' nature of algorithmic processing meaningful? The noteworthy point about MHC as a norm in the context of AWS is also that it has long been compromised in different battlefield contexts. Complex human-machine interactions are not a recent phenomenon – even the extent to which human control in a fighter jet is meaningful is questionable (Ekelhof 2019).

However, the attempts to establish MHC as an emerging norm meant to regulate AWS are difficult. Indeed, over the past four years of debate in the UN-CCW, some states, supported by civil society organisations, have advocated introducing new legal norms to prohibit fully autonomous weapons systems, while other states leave the field open in order to increase their room of manoeuvre. As discussions drag on with little substantial progress, the operational trend towards developing AI-enabled weapons systems continues and is on track to becoming established as 'the new normal' in warfare (P. W. Singer 2010). For example, in its *Unmanned Systems Integrated Roadmap 2013–2038*, the US Department of Defence sets out a concrete plan to develop and deploy weapons with ever increasing autonomous features in the air, on land, and at sea in the next 20 years (US Department of Defense 2013).

While the US strategy on autonomy is the most advanced, a majority of the top ten arms exporters, including China and Russia, are developing or planning to develop some form of AI-driven weapon systems. Media reports have repeatedly pointed to the successful inclusion of machine learning techniques in weapons systems developed by Russian arms maker Kalashnikov, coming alongside President Putin's much-publicised quote that 'whoever leads in AI will rule the world' (Busby 2018; Vincent 2017). China has reportedly made advances in developing autonomous ground vehicles (Lin and Singer 2014) and, in 2017, published an ambitiously worded government-led plan on AI with decisively increased financial expenditure (Metz 2018; Kania 2018).

The intention to regulate the practice of using force by setting norms stalls at the UN-CCW, but we highlight the importance of a reverse and likely scenario: practices shaping norms. These dynamics point to a potentially influential trajectory AWS may take towards changing what is *appropriate* when it comes to the use of force, thereby also transforming international norms governing the use of violent force.

We have already seen how the availability of drones has led to changes in how states consider using force. Here, access to drone technology appears to have made targeted killing seem an acceptable use of force for some states, thereby deviating significantly from previous understandings (Haas and Fischer 2017; Bode 2017; Warren and Bode 2014). In their usage of drone technology, states have therefore explicitly or implicitly pushed novel interpretations of key standards of international law governing the use of force, such as attribution and imminence. These practices cannot be captured with the traditional conceptual language of customary international law if they are not openly discussed or simply do not amount to its tight requirements, such as becoming 'uniform and wide-spread' in state practice or manifesting in a consistently stated belief in the applicability of a particular rule. But these practices are significant as they have arguably led to the emergence of a series of grey areas in international law in terms of shared understandings of international law governing the use of force (Bhuta et al. 2016). The resulting lack of clarity leads to a more permissive environment for using force: justifications for its use can more 'easily' be found within these increasingly elastic areas of international law.

We therefore argue that we can study how international norms regarding using AI-driven weapons systems emerge and change from the bottom-up, via deliberative and non-deliberative practices. Deliberative practices as ways of doing things can be the outcome of reflection, consideration or negotiation. Non-deliberative practices, in contrast, refer to operational and typically non-verbalised practices undertaken in the process of developing, testing and deploying autonomous technologies.

We are currently witnessing, as described above, an effort to potentially make new norms regarding AI-driven weapons technologies at the UN-CCW via deliberative practices. But at the same time, non-deliberative and non-verbalised practices are constantly undertaken as well and simultaneously shape new understandings of appropriateness. These non-deliberative practices may stand in contrast to the deliberative practices centred on attempting to formulate a (consensus) norm of meaningful human control.

This does not only have repercussions for systems currently in different stages of development and testing, but also for systems with limited AI-driven capabilities that have been in use for the past two to three decades such as cruise missiles and air defence systems. Most air defence systems already have significant autonomy in the targeting process and military aircrafts have highly automatised features (Boulanin and Verbruggen 2017). Arguably, non-deliberative practices surrounding these systems have already created an understanding of what meaningful human control is. There is, then, already a norm, in the sense of an emerging understanding of appropriateness, emanating from these practices that has not been verbally enacted or reflected on. This makes it harder to deliberatively create a new meaningful human control norm.

Friendly fire incidents involving the US Patriot system can serve as an example here. In 2003, a Patriot battery stationed in Iraq downed a British Royal Airforce Tornado that had been mistakenly identified as an Iraqi anti-radiation missile. Notably, '[t]he Patriot system is nearly autonomous, with only the final launch decision requiring human interaction' (Missile Defense Project 2018). The 2003 incident demonstrates the extent to which even a relatively simple weapons system – comprising of elements such as radar and a number of automated functions meant to assist human operators – deeply compromises an understanding of MHC where a human operator has all required information to make an independent, informed decision that might contradict technologically generated data. While humans were clearly 'in the loop' of the Patriot system, they lacked the required information to doubt the system's information competently and were therefore mislead: '[a]ccording to a summary of a report issued by a Pentagon advisory panel, Patriot missile systems used during battle in Iraq were given too much autonomy, which likely played a role in the accidental downings of friendly aircraft' (Singer

2005). This example should be seen in the context of other, well-known incidents such as the 1988 downing of Iran Air flight 655 due to a fatal failure of the human-machine interaction of the Aegis system on board the USS Vincennes or the crucial intervention of Stanislav Petrov who rightly doubted information provided by the Soviet missile defence system reporting a nuclear weapons attack (Aksenov 2013). A 2016 incident in Nagorno-Karabakh provides another example of a system with autonomous anti-radar mode used in combat: Azerbaijan reportedly used an Israeli-made Harop 'suicide drone' to attack a bus of allegedly Armenian military volunteers, killing seven (Gibbons-Neff 2016). The Harop is a loitering munition able to launch autonomous attacks.

Overall, these examples point to the importance of targeting for considering the autonomy in weapons systems. There are currently at least 154 weapons systems in use where the targeting process, comprising 'identification, tracking, prioritisation and selection of targets to, in some cases, target engagement' is supported by autonomous features (Boulanin and Verbruggen 2017, 23). The problem we emphasise here pertains not to the completion of the targeting cycle without any human intervention, but already emerges in the support functionality of autonomous features. Historical and more recent examples show that, here, human control is already often far from what we would consider as meaningful. It is noted, for example, that '[t]he S-400 Triumf, a Russian-made air defence system, can reportedly track more than 300 targets and engage with more than 36 targets simultaneously' (Boulanin and Verbruggen 2017, 37). Is it possible for a human operator to meaningfully supervise the operation of such systems?

Yet, the apparent lack/compromised form of human control is apparently considered as acceptable: neither the use of the Patriot system has been questioned in relation to fatal incidents nor is the S-400 contested for featuring an 'unacceptable' form of compromised human control. In this sense, the wider-spread usage of such air defence systems over decades has already led to new understandings of 'acceptable' MHC and human-machine interaction, triggering the emergence of new norms.

However, questions about the nature and quality of human control raised by these existing systems are not part of the ongoing discussion on AWS among states at the UN-CCW. In fact, states using automated weapons continue to actively exclude them from the debate by referring to them as 'semi-autonomous' or so-called 'legacy systems.' This omission prevents the international community from taking a closer look at whether practices of using these systems are fundamentally appropriate.

Conclusion

To conclude, we would like to come back to the key question inspiring our contribution: to what extent will AI-driven weapons systems shape and transform international norms governing the use of (violent) force?

In addressing this question, we should also remember who has agency in this process. Governments can (and should) decide how they want to guide this process rather than presenting a particular trajectory of the process as inevitable or framing technological progress of a certain kind as inevitable. This requires an explicit conversation about the values, ethics, principles and choices that should limit and guide the development, role and the prohibition of certain types of AI-driven security technologies in light of standards for appropriate human-machine interaction.

Technologies have always shaped and altered warfare and therefore how force is used and perceived (Ben-Yehuda 2013; Farrell 2005). Yet, the role that technology plays should not be conceived in deterministic terms. Rather, technology is ambivalent, making how it is used in international relations and in warfare a *political* question. We want to highlight here the 'Collingridge dilemma of control' (see Genus and Stirling 2018) that speaks of a common trade-off between knowing the impact of a given technology and the ease of influencing its social, political, and innovation trajectories. Collingridge (1980, 19) stated the following:

> Attempting to control a technology is difficult [...] because during its early stages, when it can be controlled, not enough can be known about its harmful social consequences to warrant controlling its development; but by the time these consequences are apparent, control has become costly and slow.

This describes the situation aptly that we find ourselves in regarding AI-driven weapon technologies. We are still at an initial, development stage of these technologies. Not many systems are in operation that have *significant* AI-capacities. This makes it potentially harder to assess what the precise consequences of their use in remote warfare will be. The multi-billion investments made in various military applications of AI by, for example, the USA does suggest the increasing importance and crucial future role of AI. In this context, human control is decreasing and the next generation of drones at the core of remote warfare as the practice of distance combat will incorporate more autonomous features. If technological developments proceed at this pace and the international community fails to prohibit or even regulate autonomy in weapons systems, AWS are likely to play a major role in the remote warfare of the nearer future.

At the same time, we are still very much in the stage of technological development where guidance is possible, less expensive, less difficult, and less time-consuming – which is precisely why it is so important to have these wider, critical conversations about the consequences of AI for warfare now.

References

Aksenov, Paul. 2013. 'Stanislav Petrov: The Man Who May Have Saved the World.' *BBC Russian*. September.

Amoore, Louise. 2019. 'Doubtful Algorithms: Of Machine Learning Truths and Partial Accounts.' *Theory, Culture and Society*, 36(6): 147–169.

Aradau, Claudia, and Tobias Blanke. 2018. 'Governing Others: Anomaly and the Algorithmic Subject of Security.' *European Journal of International Security,* 3(1): 1–21. https://doi.org/10.1017/eis.2017.14

Article 36. 2013. 'Killer Robots: UK Government Policy on Fully Autonomous Weapons.' http://www.article36.org/weapons-review/killer-robots-uk-government-policy-on-fully-autonomous-weapons-2/

Asaro, Peter. 2018. 'Why the World Needs to Regulate Autonomous Weapons, and Soon.' *Bulletin of the Atomic Scientists* (blog). 27 April. https://thebulletin.org/landing_article/why-the-world-needs-to-regulate-autonomous-weapons-and-soon/

Ben-Yehuda, Nachman. 2013. *Atrocity, Deviance, and Submarine Warfare*. Ann Arbor, MI: University of Michigan Press. https://doi.org/10.3998/mpub.5131732

Bhuta, Nehal, Susanne Beck, Robin Geiss, Hin-Yan Liu, and Claus Kress, eds. 2016. *Autonomous Weapons Systems: Law, Ethics, Policy*. Cambridge: Cambridge University Press.

Biegon, Rubrick, and Tom Watts. 2017. 'Defining Remote Warfare: Security Cooperation.' Oxford Research Group. https://www.oxfordresearchgroup.org.uk/Handlers/Download.ashx?IDMF=0232e573-f6d6-455e-9d34-0436925002d4

————. 2019. 'Conceptualising Remote Warfare: Security Cooperation.' Oxford Research Group.

Bode, Ingvild. 2017. '"Manifestly Failing" and "Unable or Unwilling" as Intervention Formulas: A Critical Analysis.' In *Rethinking Humanitarian Intervention in the 21st Century*, edited by Aiden Warren and Damian Grenfell. Edinburgh: Edinburgh University Press: 164–91.

———— 2019. 'Norm-Making and the Global South: Attempts to Regulate Lethal Autonomous Weapons Systems.' *Global Policy,* 10(3)*:* 359–364.

Bode, Ingvild, and Hendrik Huelss. 2018. 'Autonomous Weapons Systems and Changing Norms in International Relations.' *Review of International Studies,* 44(3): 393–413.

Boulanin, Vincent, and Maaike Verbruggen. 2017. 'Mapping the Development of Autonomy in Weapons Systems.' Stockholm: Stockholm International Peace Research Institute. https://www.sipri.org/sites/default/files/2017-11/siprireport_mapping_the_development_of_autonomy_in_weapon_systems_1117_1.pdf

Busby, Mattha. 2018. 'Killer Robots: Pressure Builds for Ban as Governments Meet.' *The Guardian*, 9 April 9. sec. Technology. https://www.theguardian.com/technology/2018/apr/09/killer-robots-pressure-builds-for-ban-as-governments-meet

Casey-Maslen, Stuart. 2012. 'Pandora's Box? Drone Strikes under Jus Ad Bellum, Jus in Bello, and International Human Rights Law.' *International Review of the Red Cross,* 94(886): 597–625.

Cavallaro, James, Stephan Sonnenberg, and Sarah Knuckey. 2012. 'Living Under Drones: Death, Injury and Trauma to Civilians from US Drone Practices in Pakistan.' International Human Rights and Conflict Resolution Clinic, Stanford Law School/NYU School of Law, Global Justice Clinic. https://law.stanford.edu/publications/living-under-drones-death-injury-and-trauma-to-civilians-from-us-drone-practices-in-pakistan/

Cole, David. 2014. 'We Kill People Based on Metadata.' *The New York Review of Books* (blog). 10 May. https://www.nybooks.com/daily/2014/05/10/we-kill-people-based-metadata/

Collingridge, David. 1980. *The Social Control of Technology*. London: Frances Pinter.

Der Derian, James. 2009. *Virtuous War: Mapping the Military-Industrial-Media-Entertainment Network*. 2nd ed. New York: Routledge.

Ekelhof, Merel. 2019. 'Moving Beyond Semantics on Autonomous Weapons: Meaningful Human Control in Operation.' *Global Policy*, 10(3): 343–348. https://doi.org/10.1111/1758-5899.12665

Farrell, Theo. 2005. *The Norms of War: Cultural Beliefs and Modern Conflict*. Boulder: Lynne Rienner Publishers.

Fleischman, William M. 2015. 'Just Say "No!" To Lethal Autonomous Robotic Weapons.' *Journal of Information, Communication and Ethics in Society*, 13(3/4): 299–313.

Genus, Audley, and Andy Stirling. 2018. 'Collingridge and the Dilemma of Control: Towards Responsible and Accountable Innovation.' *Research Policy*, 47(1): 61–69.

Gibbons-Neff, Thomas. 2016. 'Israeli-Made Kamikaze Drone Spotted in Nagorno-Karabakh Conflict.' *The Washington Post*. 5 April. https://www.washingtonpost.com/news/checkpoint/wp/2016/04/05/israeli-made-kamikaze-drone-spotted-in-nagorno-karabakh-conflict/?utm_term=.6acc4522477c

Gregory, Thomas. 2015. 'Drones, Targeted Killings, and the Limitations of International Law.' *International Political Sociology*, 9(3): 197–212.

Gusterson, Hugh. 2016. *Drone: Remote Control Warfare*. Cambridge, MA/London: MIT Press.

Haas, Michael Carl, and Sophie-Charlotte Fischer. 2017. 'The Evolution of Targeted Killing Practices: Autonomous Weapons, Future Conflict, and the International Order.' *Contemporary Security Policy*, 38(2): 281–306.

Hall, Abigail R., and Christopher J. Coyne. 2013. 'The Political Economy of Drones.' *Defence and Peace Economics*, 25(5): 445–60.

ICRC. 2016. 'Views of the International Committee of the Red Cross (ICRC) on Autonomous Weapon Systems.' https://www.icrc.org/en/document/views-icrc-autonomous-weapon-system

Kaempf, Sebastian. 2018. *Saving Soldiers or Civilians? Casualty-Aversion versus Civilian Protection in Asymmetric Conflicts*. Cambridge: Cambridge University Press.

Kahn, Paul W., 2002. 'The Paradox of Riskless Warfare.' *Philosophy and Public Policy Quarterly,* 22(3): 2–8.

Kania, Elsa. 2018. 'China's AI Agenda Advances.' *The Diplomat*. 14 February. https://thediplomat.com/2018/02/chinas-ai-agenda-advances/

Kilcullen, David, and Andrew McDonald Exum. 2009. 'Death From Above, Outrage Down Below.' *The New York Times*. 17 May.

Lin, Jeffrey, and Peter W. Singer. 2014. 'Chinese Autonomous Tanks: Driving Themselves to a Battlefield Near You?' *Popular Science*. 7 October. https://www.popsci.com/blog-network/eastern-arsenal/chinese-autonomous-tanks-driving-themselves-battlefield-near-you

Mandel, Robert. 2004. *Security, Strategy, and the Quest for Bloodless War*. Boulder, CO: Lynne Rienner Publishers.

Metz, Cade. 2018. 'As China Marches Forward on A.I., the White House Is Silent.' *The New York Times*. 12 February. sec. Technology. https://www.nytimes.com/2018/02/12/technology/china-trump-artificial-intelligence.html

Missile Defense Project. 2018. 'Patriot.' Missile Threat. https://missilethreat.csis.org/system/patriot/

Morozov, Evgeny. 2014. *To Save Everything, Click Here: Technology, Solutionism and the Urge to Fix Problems That Don't Exist*. London: Penguin Books.

Oudes, Cor, and Wim Zwijnenburg. 2011. 'Does Unmanned Make Unacceptable? Exploring the Debate on Using Drones and Robots in Warfare.' IKV Pax Christi.

Restrepo, Daniel. 2019. 'Naked Soldiers, Naked Terrorists, and the Justifiability of Drone Warfare:' *Social Theory and Practice,* 45(1): 103–26.

Roff, Heather M., and Richard Moyes. 2016. 'Meaningful Human Control, Artificial Intelligence and Autonomous Weapons. Briefing Paper Prepared for the Informal Meeting of Experts on Lethal Autonomous Weapons Systems. UN Convention on Certain Conventional Weapons.'

Sauer, Frank, and Niklas Schörnig. 2012. 'Killer Drones: The 'Silver Bullet' of Democratic Warfare?' *Security Dialogue,* 43(4): 363–80.

Scheipers, Sibylle, and Bernd Greiner, eds. 2014. *Heroism and the Changing Character of War: Toward Post-Heroic Warfare?* Houndmills: Palgrave Macmillan.

Schwarz, Elke. 2016. 'Prescription Drones: On the Techno-Biopolitical Regimes of Contemporary 'ethical Killing.' *Security Dialogue,* 47(1): 59–75.

Sharkey, Noel. 2008. 'The Ethical Frontiers of Robotics.' *Science,* 322(5909): 1800–1801.

Sharkey, Noel. 2010. 'Saying 'No!' To Lethal Autonomous Targeting.' *Journal of Military Ethics,* 9(4): 369–83.

Singer, Jeremy. 2005. 'Report Cites Patriot Autonomy as a Factor in Friendly Fire Incidents.' SpaceNews.Com. 14 March. https://spacenews.com/report-cites-patriot-autonomy-factor-friendly-fire-incidents/

Singer, Peter W., 2010. *Wired for War. The Robotics Revolution and Conflict in the 21st Century.* New York: Penguin.

Sparrow, Robert. 2007. 'Killer Robots.' *Journal of Applied Philosophy,* 24(1): 62–77.

Strawser, Bradley Jay. 2010. 'Moral Predators: The Duty to Employ Uninhabited Aerial Vehicles.' *Journal of Military Ethics,* 9(4): 342–68.

The Associated Press. 2018. 'Tens of Thousands of Russian Troops Have Fought in Syria since 2015.' Haaretz. 22 August. https://www.haaretz.com/middle-east-news/syria/tens-of-thousands-of-russian-troops-have-fought-in-syria-since-2015-1.6409649

The Intercept. 2015. 'SKYNET: Courier Detection via Machine Learning - The Intercept.' 2015.
https://theintercept.com/document/2015/05/08/skynet-courier/

United States. 2018. 'Human-Machine Interaction in the Deveopment, Deployment and Use of Emerging Technologies in the Area of Lethal Autonomous Weapons Systems. UN Document CCW/GGE.2/2018/WP.4.'
https://www.unog.ch/80256EDD006B8954/(httpAssets)/
D1A2BA4B7B71D29FC12582F6004386EF/$file/2018_GGE+LAWS_August_
Working+Paper_US.pdf

US Department of Defense. 2013. 'Unmanned Systems Integrated Roadmap: FY2013-2038.'
https://info.publicintelligence.net/DoD-UnmannedRoadmap-2013.pdf

Vincent, James. 2017. 'Putin Says the Nation That Leads in AI 'Will Be the Ruler of the World.'' The Verge. 4 September. https://www.theverge.
com/2017/9/4/16251226/russia-ai-putin-rule-the-world

Walsh, James Igoe, and Marcus Schulzke. 2018. *Drones and Support for the Use of Force*. Ann Arbor: University of Michigan Press.

Warren, Aiden, and Ingvild Bode. 2014. *Governing the Use-of-Force in International Relations. The Post-9/11 US Challenge on International Law*. Basingstoke: Palgrave Macmillan.

———. 2015. 'Altering the Playing Field: The US Redefinition of the Use-of-Force.' *Contemporary Security Policy,* 36 (2): 174–99. https://doi.org/10.1080/
13523260.2015.1061768

14

Conclusion: Remote Warfare in an Age of Distancing and 'Great Powers'

ALASDAIR MCKAY

In the introduction, it was stated that the main goals of this edited volume were to start filling the gaps in our understanding of remote warfare, challenge the dominant narratives surrounding its use and subject the practice to greater scrutiny. Through reading this book, readers will hopefully be left with a better comprehension of remote warfare than when they opened to the first page. Moreover, the three interconnected core themes of this book, revisited below, have challenged the conventional wisdom and exposed some of remote warfare's serious problems.

Firstly, though it can yield some short-term tactical successes, remote warfare is not a silver bullet solution to the deep-set political problems in conflict-affected states. In fact, it can damage peace and stability in states where it is used. Several chapters have shown how the use of remote warfare can exacerbate the drivers of conflict. This has been true whether it is remote warfare in Syria, as Sinan Hatahet's chapter demonstrated, Libya, discussed in the editors' conceptual introduction, or the Sahel, as explored by Delina Goxho.

Secondly, despite being presented at 'precise', 'surgical' and even 'humane', remote military engagements often do cause significant harm to civilians. Remote warfare does minimise the risks to a state's own soldiers, but in doing so, it shifts the burdens of warfare onto civilians. As Baraa Shiban and Camilla Molyneux's chapter on Yemen illustrated, the harm inflicted on civilians through remote warfare is not limited simply to deaths. It and can also have significant economic, educational, and mental health implications for impacted communities. Civilian harm in remote warfare is also closely linked to instability. As Daniel Mahanty argued in part of his chapter on security cooperation, civilian harm and human rights violations committed by partners can counteract peacebuilding initiatives. It can also erode the public's trust in the legitimacy of the partner state and increase the number of the disaffected who may turn to violence in response to state-sponsored abuse.

Finally, remote warfare has significant socio-political impacts on the states that practice it. The secrecy surrounding the use of remote warfare is potentially having a corrosive impact on democratic norms. As the chapter by Christopher Kinsey and Helene Olsen noted on private militaries, there is a danger that the lack of debate on their use could create a democratic deficit, where accountability, transparency, and even public consent are either ignored or quietly marginalised. According to Malte Riemann and Norma Rossi's chapter, outsourcing the burdens of warfare has had a deeper effect of reshaping modes of remembrance, duty, and sacrifice in states. This has subsequently made war appear less visible within democratic societies. Jolle Demmers and Lauren Gould warn that there is a danger that in the long term, with the removal of warfare from visibility and scrutiny, Western liberal democracies could become more violent.

As noted in the introduction there are limitations to what can be covered in any book and there are always areas left unexplored. Though the articles have been deep in their analyses, this volume has only scratched the surface of the scale and scope of remote warfare. As such, this concluding chapter examines some of the different thematic areas that could be explored in future research on remote warfare. But first the chapter discusses the important question of whether remote warfare will remain the norm for states, particularly given the rise of 'great power competition' and the COVID-19 pandemic. These developments have yielded important questions regarding the future of remote warfare.

Is remote warfare here to stay?

Of late, there has been much talk from International Relations scholarship, think tanks, the defence community and politicians that that we once again live in a time of 'great power competition' (Dueck 2017; Kaufmann 2019; Elbridge and Mitchell 2020; Mahnken 2020). The crux of the idea is that there has been a shift away from global hegemony and towards a world where the US, China and Russia compete for strategic influence, trade and investment dominance, and world leader status in the development and regulation of new technologies (O'Rourke 2020). For states, this has meant that near-peer competition has become the main strategic priority, rather than counterterrorism. The 2018 National Defence Strategy, for example, outlines that: 'Inter-state strategic competition, not terrorism, is now the primary concern in US national security' (United States Department of Defence 2018, 1).

This grand narrative has been gathering momentum for a while. At the start of the 1990s, John Mearsheimer (1990, 5–6) opined, 'the bipolar structure that has characterised Europe since the end of World War II is replaced by a multipolar structure.' Since then, various writers have examined the military revival of Russia (Trenin 2016; Renz 2017), the economic and military rise of China (Kristof 1992; Overholt 1994; Buzan 2010) and the implications of all this for international security.

However, developments over the past decade have been seen to strengthen the validity of this narrative. In 2014, driven by numerous factors, Vladimir Putin invaded Ukraine and 'annexed' Crimea which sent alarm bells ringing in the West and gave NATO a renewed purpose. Since then, Russia has expanded its presence in many parts of the world through arms sales, an undeclared, but seemingly significant, presence of mercenaries and special forces abroad, as well as capacity-building programmes for local forces (Watson and Karlshøj-Pedersen 2019). China's 'aggressive' trade activity (Lukin 2019), investment in defence technologies (Maizland 2020) and human rights abuses (Human Rights Watch 2019) have also raised concerns in the West. Since becoming China's paramount leader in 2012, President Xi Jinping has been accused of pursuing an ambitious, nationalistic agenda abroad, evidenced by Chinese claims to disputed territory in the South China Sea (Nouwens 2020), face-offs with India in the Galwan Valley (Wu and Myers 2020) and behaviour towards Taiwan (Ford and Gewirtz 2020).

There are also domestic drivers behind this rise of 'great power competition.' Though hostilities between powers pre-date the rise of 'strongman politics', this development is likely to be a significant factor. As Lawrence Freedman (2020) recently noted:

> In the age of Trump, Xi, and Putin, it is hard to take seriously the idea that domestic affairs have only a trivial effect on the logic of great power practice. Moreover, domestic affairs not only help explain strategic choices, in terms of identifying interests and making provisions for warfare, but also what the powers have on offer. The way they govern themselves and arrange their social and economic affairs is part of the influence they exert.

Since coming to power in 2016, President Trump has made this 'great power competition' grand narrative the centrepiece of US defence and security thinking (Rachman 2019). The Obama administration were certainly concerned about Russia and China as parts of the 2015 National Security Strategy illustrated (White House 2015). But the 2017 National Security Strategy (White House 2017, 2) represented a formal announcement of this shift in global relations: 'After being dismissed as a phenomenon of an earlier century [...] great power competition returned.' In more recent comments, Defense Secretary Mark Esper outlined US strategic priorities:

> For the United States, our long-term challenges, China, No. 1, and Russia, No. 2. And what we see happening out there is a China that continues to grow its military strength, its economic power, its commercial activity, and it's doing so, in many ways, illicitly — or it's using the international rules-based order against us to continue this growth, to acquire technology, and to do the things that really undermine our [and our allies'] sovereignty, that undermine the rule of law, that really question [Beijing's] commitment to human rights. (quoted in Kristian 2020)

The US National Security Strategy also identifies other rising powers, such as Iran and North Korea as strategic concerns, and their attempts to 'destabilize regions, threaten Americans and our allies, and brutalize their own people' (DOD 2017, 15).

This rise of the 'great power competition' narrative has created new uncertainties for international security, not least for the use of remote warfare as a tactical tool for states. But there are reasons to be doubtful that it will mark the end for remote warfare or a return to large-scale interventions.

In the 'great power competition' era, states such as the US will rely heavily on partnerships. As Watts, Biegon and Mahanty noted in their chapters, security cooperation will likely remain an important tool in the American foreign policy. This will likely be true in the case of its allies too. Several countries are considering following a light-footprint strategy of 'persistent engagement', where a state 'maintains a presence in a country, with few troops, and work with regional and local partners to try and build influence and knowledge' (Watson 2020b).

Recent trends also show that states continue to have a strong strategic interest in confronting adversaries' armed forces off the open battlefield, operating in the grey zone and under the threshold of full, state-on-state conflict (Knowles and Watson 2018, 5–6). Remote approaches are essentially ways for states to avoid the economic and political risks of direct confrontation. The assassination of General Qasem Soleimani earlier this year by a US armed drone strike is an example of how remote warfare has been used to avoid direct confrontation, as are Iran's use of proxies in the Middle East. Both nations have sought to avoid directly fighting, but in doing so they have shifted the risk onto local civilians in the areas they are engaged.

In the case of Russia, it is also constrained economically and by manpower limitations. These realities led the RAND Corporation to conclude:

> There is no indication that Russia is seeking a large-scale conflict with a near-peer or peer competitor, and indeed it appears Russian leaders understand the disadvantages Russia faces in the event of a prolonged conflict with an adversary like NATO. (Boston and Massicot 2018)

So far, Putin's approach to the West has largely taken the form of cyber operations, disinformation campaigns and targeted assassinations (see Thomas 2014; Connell and Vogler 2017; Mejias and Vokuev 2017; Stengel 2019; Splidsboel Hansen 2017). Reasonably competent at working 'on the cheap', Putin has also used limited remote military interventions as a broader foreign policy tool. This is likely to continue. As such, it is more probable to find US or UK troops in future confrontation with states like Russia, via its military contractors or special forces, in somewhere like Syria, rather than in a conventional war in Eastern Europe (Knowles and Watson 2018, 6). There is certainly a precedent for this. In February 2018, it was reported that US Special Forces clashed with Russian security contractors, working with Syrian forces, as part of a four-hour long firefight in eastern Syria (Gibbons-Neff 2018). The heavy Russian losses from this engagement, reportedly 200 troops (Ibid.), and the reputational damage may arguably make the Kremlin more hesitant about repeating this type of event. But this does not rule out skirmishes of a similar nature reoccurring.

In this sense, military engagement between 'great powers' and their allies is more likely to take the form of remote warfare or at least display elements of it. This presents a number of challenges to the transparency and accountability – and many of the dangers discussed throughout the book are likely to continue.

The recent COVID-19 outbreak, one of the largest global pandemics in living memory, has undoubtably increased tensions between China and the West.

This volume was being finalised during the early stages of the outbreak. In just a few months, the spread of the virus has ground many countries across the world to a standstill and chronically impacted their economies and social routines. Estimates put the death toll so far at over one million (World Health Organisation 2020) and the cost to the global economy at between £4.7–£7.1 trillion (Asia Development Bank 2020). The origins of the virus in Wuhan, a city in central China, and the rapid global spread which followed has led some to blame China for the impact of the virus. However, this growing Sino–Western rivalry still remains below the threshold of major war and is unlikely to change. Hostilities will likely take the form of sanctions, cyber conflicts and potentially proxy engagements.

In their response to COVID-19, some governments have taken a heavily securitised approach and, in some cases, exploited the situation to consolidate power (Roth 2020; Lamond 2020). This has seen state security agencies abuse their positions of authority and act outside the rule of law, often engaging in overly aggressive measures towards civilians (Brooks 2020). There is a danger that these actions could damage the relationship between the state and its people, help foster grievances, push alienated civilians towards to extremist groups and contribute to more violence in the long run (Watson 2020a).

There have been several warnings that non-state armed groups are attempting to exploit the disorder created by the pandemic in certain states. In Iraq, the Islamic State issued instructions to supporters regarding the virus and began to intensify its various attacks all over the Middle East and other regions (Abu Haneyeh 2020). In the Sahel, another area seen as a prominent battleground for jihadist groups, al-Qaeda affiliates and the Islamic State in the Greater Sahara have also attempted to make gains from the outbreak and carried out attacks against military positions, UN peacekeepers and civilian populations (Berger 2020). An analysis by the Center for Strategic and International Studies, using the Armed Conflict Location and Event Data Project database, noted that violent attacks in Sub-Saharan Africa's conflict hotspots rose by 37 percent in the early months of 2020 when the virus was spreading in the region (Colombo and Harris 2020). Yet even before the pandemic, there were several warnings about the resurgence of Islamic State and growing presence of al-Qaeda, not only in Africa, but also the Middle East and South East Asia (Felbab-Brown 2019; Hassan 2019; Joffé 2018; Lefèvre 2018; Clarke 2019; Jones Harrington 2018). The bottom line is that non-state armed groups are likely to remain a threat for some time.

As such, it is highly probable that remote warfare will be the preferred method used to counter them because it is seen by them as low risk and relatively cheap. Many analysts certainly feel this to be the case with the UK. In an expert roundtable hosted by Oxford Research Group in early 2020, the participants indicated that the economic and political climate in the country would mean that the UK is likely to continue to take a remote approach to military engagements in the future. In recent years, the UK has seen its markets impacted by the uncertainty over Brexit, the economy crippled because of the COVID-19 responses, and the Government under pressure to reduce spending (Watson 2020b). In a general sense, the military, political and economic constraints that initially led to the dominance of remote warfare are still present and will likely be exacerbated (see Chalmers and Jessett 2020). Despite a changing global landscape, remote warfare is therefore likely to continue to define the approach of many states, making critical enquiry on the subject matter all the more important.

Some future directions of research

Listening to and Including local voices

A common narrative of this book is that while remote warfare may be 'remote' from Western perspectives, it is part of the everyday reality for some communities in Africa, the Middle East, Asia and elsewhere. It has significant impacts on civilian populations and much of this remains underreported. But as the chapter by Shiban and Molyneux on Yemen highlighted, conceptualisations of civilian harm in remote warfare need to move beyond civilian deaths and injuries to broader understandings of its effect on societies.

These realities make it important to find and amplify the voices of the communities in states where remote warfare operations are conducted. Work by investigative journalists, academics and NGOs has been invaluable in bringing these currently marginalised voices into clearer focus (see Watling and Shabibi 2018; Pargeter 2017). But this remains a very limited and restricted research exercise. There are good reasons for this. Field research in the terrains of remote warfare is both costly and dangerous (see Bliesemann de Guevara and Kurowska 2020). Nonetheless, greater inclusion of local populations' perspectives on how they perceive the use remote warfare in their communities would undoubtably improve understandings of the phenomenon and give a voice to those who have largely been ignored in discussions.

Getting local voices heard does not necessarily have to be done by field work. The internet holds huge potential to offer a platform to marginalised voices. A future online edited volume on remote warfare could be based around commissioning chapters from individuals and groups in theatres where operations have taken place.

Examining non-Western approaches

This book has been largely concerned with critiquing Western states' use of remote warfare, particularly the engagement of the US and UK. Though some chapters did certainly explore the non-Western dynamics to remote warfare, there is nonetheless a greater weighting towards Western approaches. Western states rely heavily on remote warfare and so it makes sense for researchers in Western democracies to focus their attentions on the activities of their own governments and militaries because there is a greater chance of stimulating change. Moreover, the general lack of debate on remote warfare in the West makes it essential for researchers to lead the way in raising awareness of these issues.

Nevertheless, expanding the scope of the case studies to explore non-Western approaches to remote warfare could be a fruitful avenue for scholars to explore. There is, of course, no shortage of literature exploring the use of remote approaches to fighting by the likes of Russia, Iran, China or the Gulf States (Mumford 2013; Berti and Guzansky 2015; Renz 2016; Chivvas 2017; Fridman 2018; Kuzio and D'Anieri 2018, 25–61; Fabian 2019; Krieg 2018; Krieg and Rickli 2019). There are also several accounts on the use of remote tactics by developing states, particularly those in Africa (Abbink 2003; Tubiana and Walmsley 2008; Craig 2012; Tamm 2014; Isaacs-Martin 2015; 2018; Krieg and Rickli 2018; Tapscott 2019; International Crisis Group 2020). Nonetheless, a comparison between democratic and less democratic states' experiences of remote warfare would be a worthwhile pursuit. It may help researchers to understand the differences and similarities between how states use remote approaches. A particularly interesting question to address on this topic could be whether there is a relationship between regime type and remote warfare and, if so, what the drivers behind this are.[1] As remote warfare is likely to be a tool used by states for some time, a greater focus on how approaches to remote warfare differ across the globe may become even essential in the future.

Watchful eyes on technology

The technological tools used in remote warfare today, such as drones, will still

---

[1]   For some initial data collection on this, see a presentation by Yvonni Efstathiou on regime type and the use of non-state armed groups: https://www.oxfordresearchgroup. org.uk/event-podcast-the-oversight-and-accountability-of-remote-warfare

be present in the short-term and will be an important area of future research. Scholars and researchers will continue to raise awareness on how the use of such technology impacts civilians on the ground and its broader ramifications, particularly its contribution to greater radicalisation and subsequent instability (see Saeed et al. 2019). But as the chapters noted there are concerns that technological advances in defence are outpacing legal and moral frameworks both domestically and internationally.

The increasing flow of global data, which is driven by new information technologies, is one example of this. As Julian Richards noted in his chapter on intelligence sharing, there is a risk that highly complex and integrated intelligence systems, sharing ever more industrial-scale amounts of data, could enable abuses of intelligence by states. In his chapter Richards notes that there are public fears in Western democracies about a creep towards a global 'surveillance society' and that intelligence sharing with authoritarian regime could contribute to greater human rights abuses.

On the same general theme, Jennifer Gibson's chapter highlighted the dangers of data-driven approaches to targeted killing through armed drone strikes, and the challenges this activity poses to international law. As Gibson argued, in places like Yemen life and death decisions are being made based on loose collections of data assembled by algorithms with limited intelligence on the ground. This raises difficult questions about whether technology helps or hinders the processes that lead to pilots launching deadly drone strikes.

Joseph Chapa, whose research involved interviews with armed drone pilots, came to a more optimistic conclusion about how the distance in remote warfare, enabled by technology, impacts pilots' judgement. In his chapter Chapa argued that drone technology actually enables pilots to exercise human judgement when making life and death decisions. Nevertheless, Chapa did also point to the potential dangers presented by emerging technologies like artificial intelligence (AI) to this process.

Indeed, perhaps the greatest anxiety surrounding future developments in military technology concerns the dawn of autonomous weapons systems (AWS) and AI (see Scharre 2014, 2019; Sharkey 2017; Schwarz 2018). This is an emerging global phenomenon, with global military spending on AWS and AI projected to reach $16 and $18 billon respectively by 2025 (Sander and Meldon 2014). A growing number of states and non-governmental organisations are appealing to the international community for regulation of or even bans on AWS (Cummings 2017, 2). Certainly, there are valid ethical concerns about AWS. As Ingvild Bode and Hendrik Huelss' chapter highlighted, these technologies could challenge the existing norms governing the use of force due to the effect they may have on human judgement. This could have huge impacts on civilians in warfare. According to their chapter, 'the legal definition of who is a civilian and who is a combatant is not written in a way that could be easily programmed into AI, and machines lack the situational awareness and ability to infer things necessary to make this decision.'

However, some are more optimistic about AI, particularly concerning its relationship with civilian harm. Though researchers across various disciplines are cautious about the growth of this technology, they believe that, if used under the right conditions, such systems have potentially more 'positive' uses (for a good overview of the key debates see ICRC 2019). In terms of its impact on warfare and civilian harm, Larry Lewis, director of the Centre for Autonomy and Artificial Intelligence, has argued that the proper use of machine learning algorithms can help minimise civilian casualties during armed conflict:

> While the history of warfare is replete with examples of technology being used to kill and maim more people more efficiently, technology can also reduce those tragic costs of war. For example, precision-guided and small-sized munitions can limit so-called collateral damage, the killing and maiming of civilians and other non-combatants. (Lewis 2018)

Going forward, more open debate, dialogue and the circulation of accurate information will be crucial. This will mean that there is a shared understanding of risks and ways to better promote safety for the military applications of technology. The lack of discussion and progress among UN member states on this subject shows (see Haner and Garcia 2019) that the international community has a lot of catching up to do on this issue.

Looking forward: the value of intellectual pluralism

As the introduction noted, in 2019 an event was co-organised by Oxford Research Group and the University of Kent which brought together stakeholders from various academic disciplines, the NGO community, civil society, and the military to discuss remote warfare. The event showed how important engagement across professional sectors can be as both a learning experience and in moving conversation forward (Watts and Biegon 2019). The conference saw those in the military and NGO sectors, communities that might not normally share platforms, exchange their experiences of remote warfare. This book has captured some of that diversity by shedding light on the key debates permeating the use of remote warfare.

As this volume has shown, remote warfare affects many sectors of societies both at home and abroad. It is not simply a military matter, but rather a highly social one. Inclusive, open and diverse dialogue and debate between stakeholders involved in remote warfare, then, is vital if scholarship is to continue to grow. If researchers fail to reach beyond professional silos and work collaboratively, it risks creating a stale discursive environment where research clusters fall into circular discussions in their own echo chambers. The chapters in the book have shown that the use of remote warfare has several significant problems and there can only be progress towards resolving them through discussions between communities. This book, then, represents part of the beginning of this process, not its end.

References

Abbink, John. 2003. 'Ethiopia—Eritrea: Proxy wars and prospects of peace in the horn of Africa.' *Journal of Contemporary African Studies*, 21(3), 407–426.

Abu Haneyeh, Hassan. 2020. 'How COVID-19 Facilitated the Rebirth of Global Jihadism.' Pulitzer Center. 3 June. https://pulitzercenter.org/reporting/how-covid-19-facilitated-rebirth-global-jihadism

Asian Development Bank. 2020. 'Updated Assessment of the Potential Economic Impact of COVID-19.' ADB Briefs, No. 133. May.

Berger, Flore, 2020. 'Business as usual for jihadists in the Sahel, despite pandemic.' IISS Analysis. 29 April. https://www.iiss.org/blogs/analysis/2020/04/csdp-jihadism-in-the-sahel

Berti, Benedetta and Guzansky, Yoel. 2014. 'Saudi Arabia's Foreign Policy on Iran and the Proxy War in Syria: Toward a New Chapter?' *Israel Journal of Foreign Affairs*, 8(3): 25–34.

Bliesemann de Guevara, Berit, and Bøås, Morten, eds. 2020. *Doing Fieldwork in Areas of International Intervention: A Guide to Research in Violent and Closed Contexts. Spaces of Peace, Security and Development.* Bristol University Press.

Boston, Scott, and Dara Massicot. 2017. 'The Russian Way of Warfare: A Primer.' Santa Monica, CA: RAND Corporation. https://www.rand.org/pubs/perspectives/PE231.html

Brooks, Lewis. 2020. 'The role of the security sector in COVID-19 response An opportunity to "build back better"?' Saferworld.

Buzan, Barry. 2010. 'China in international society: Is "peaceful rise" possible?.' *The Chinese Journal of International Politics*, 3(1): 5–36.

Chalmers, Malcolm and Will Jessett. 2020. 'Will Defence and the Integrated Review: A Testing Time.' Whitehall Reports. RUSI. 26 March.

Chivvis, Christopher S., 2017. 'Understanding Russian "Hybrid Warfare": And What Can Be Done About It: Addendum.' Santa Monica, CA: RAND Corporation. https://www.rand.org/pubs/testimonies/CT468z1.html

Clarke, Colin. P., 2019. *After the Caliphate: The Islamic State and the Future Terrorist Diaspora*. John Wiley and Sons.

Columbo, Emilia, and Marielle Harris. 2020. 'Extremist groups stepping up operations during the covid-19 outbreak in sub-Saharan Africa.' Centre for Strategic and International Studies. 1 May.

Craig, Dylan. 2012. *Proxy war by African states, 1950–2010*. American University.

Colby, Elbridge A., and A. Wess Mitchell. 2020. 'The Age of Great-Power Competition: How the Trump Administration Refashioned American Strategy.' *Foreign Affairs,* 99: 118. January/February.

Connell, Michael, and Sarah Vogler. 2017. 'Russia's Approach to Cyber Warfare (1Rev).' Center for Naval Analyses Arlington United States.

Cummings, Missy. 2017. 'Artificial intelligence and the future of warfare.' London: Chatham House.

Dueck, Colin. 2017. 'An Era of Great-Power Leaders.' *The National Interest*. 7 November.

Efstathiou, Yvonni. 'Regime Type and Remote Warfare: A Principal Analysis of International War'. Oxford Research Group. https://www.oxfordresearchgroup.org.uk/event-podcast-the-oversight-and-accountability-of-remote-warfare

Fabian, Sandor. 2019. 'The Russian hybrid warfare strategy – neither Russian nor strategy.' *Defense and Security Analysis,* 35(3): 308–325.

Felbab-Brown, Vanda. 2017. 'Afghanistan's terrorism resurgence: Al-Qaida, ISIS, and beyond.' Testimony Transcript Subcommittee on Terrorism, Non-proliferation, and Trade of the House Foreign Affairs Committee. Brookings Institute.

Ford, Lindsey W., and Julian Gewirtz. 2020. 'China's Post-Coronavirus Aggression Is Reshaping Asia.' *Foreign Policy.* 18 June. https://foreignpolicy.com/2020/06/18/china-india-aggression-asia-alliances/

Freedman, Lawrence. 2020. 'Who Wants to Be a Great Power?' *PRISM,* 8(4): 2–15.

Fridman, Ofer. 2018. *Russian 'Hybrid Warfare': Resurgence and Politicisation.* New York: Oxford University Press.

Friedman, Uri. 2019. 'The New Concept Everyone in Washington Is Talking About.' *The Atlantic,* 6 August.

Gibbons-Neff, Thomas. 2018. 'How a 4-hour battle between Russian mercenaries and US commandos unfolded in Syria.' *New York Times.* 24 May.

Haner, Justine, and Denise Garcia. 2019. 'The Artificial Intelligence Arms Race: Trends and World Leaders in Autonomous Weapons Development.' *Global Policy*, 10(3): 331–337.

Hassan, Hassan. 2018. 'ISIS Is Ready for a Resurgence.' *The Atlantic.* 26 August.

Human Rights Watch. 2019. *China's Global Threat to Human Rights.* https://www.hrw.org/world-report/2020/country-chapters/global

International Committee on the Red Cross. 2019. 'Artificial Intelligence and Armed Conflict', Humanitarian Law and Policy Blog. ICRC. https://blogs.icrc.org/law-and-policy/category/special-themes/artificial-intelligence/

International Crisis Group. 2020. 'Averting Proxy Wars in the Eastern DR Congo and Great Lakes.' Briefing 150 / Africa. 23 January.

Isaacs-Martin, Wendy. 2015. 'The motivations of warlords and the role of militias in the Central African Republic.' *Conflict Trends*, 2015(4): 26–32.

Joffé, George. 2016. 'The fateful phoenix: the revival of Al-Qa'ida in Iraq.' *Small Wars and Insurgencies*, 27(1): 1–21.

Jones, Seth G., Charles Vallee, Danika Newlee, Nicholas Harrington, Clayton Sharb, and Hannah Byrne. 2018. 'The Evolution of the Salafi-Jihadist Threat.' Center for Strategic International Studies. 20 November.

Knowles, Emily, and Abigail Watson. 2018. *Remote Warfare: Lessons Learned from Contemporary Theatres.* Oxford Research Group.

Krieg, Andreas. 2017. *Defining Remote Warfare: The Rise of the Private Military and Security Industry*. Oxford Research Group.

Krieg, Andreas and Jean-Marc Rickli. 2018. 'Surrogate warfare: the art of war in the 21st century?' *Defence Studies*, 18(2): 113–130.

————. 2019. *Surrogate Warfare: The Transformation of War in the Twenty-First Century*. Washington: Georgetown University Press.

Kristian, Bonnie. 2020. 'Esper's dark vision for US-China conflict makes war more likely.' *DefenseNews*. 19 March.

Kristof, Nicholas. 1992. 'The rise of China.' *Foreign Affairs*, 72: 59. November/December.

Kuzio, Taras, and Paul D'Anieri. 2018. *Sources of Russia's great power politics: Ukraine and the challenge to the European order*. Bristol: E-International Relations.

Lamond, James. 2020. 'Authoritarian Regimes Seek To Take Advantage of the Coronavirus Pandemic.' Center for American Progress. 6 April.

Lefèvre, Raphaël. 2018. 'The resurgence of Al-Qaeda in the Islamic Maghrib.' *The Journal of North African Studies*, 23(1-2): 278–281.

Lewis, Larry. 2018. 'AI for Good in War; Beyond Google's "Don't Be Evil:"' *Breaking Defence*. 29 June.

Lukin, Alexander. 2019. 'The US–China Trade War and China's Strategic Future.' *Survival*, 61(1): 23–50.

Mahnken, Thomas. 2020. *Forging the Tools of 21st Century Great Power Competition*. Center for Strategic and Budgetary Assessments.

Maizland, Lindsay. 2020. 'China's Modernizing Military.' Council on Foreign Relations. 5 February.
https://www.cfr.org/backgrounder/chinas-modernizing-military

Mearsheimer, John. 1990. 'Back to the Future: Instability in Europe after the Cold War.' *International Security*, 15(1): 5–56.

Mejias, Ulises, and Nikolai Vokuev. 2017. 'Disinformation and the media: the case of Russia and Ukraine.' *Media, Culture and Society*, 39(7): 1027–1042.

Mumford, Andrew. 2013. *Proxy warfare*. Cambridge: Polity.

Nouwens, Veerle. 2020. 'Who Guards The "Maritime Silk Road"?' *War On The Rocks.* 27 June.
https://warontherocks.com/2020/06/who-guards-the-maritime-silk-road/#:~:text=The%20Zhongjun%20Junhong%20Security%20Group's,reconnaissance%20forces%2C%20and%20marine%20corps

O'Rourke, Ronald. 2020. *Renewed Great Power Competition: Implications for Defense: Issues for Congress*. Congressional Research Service. 29 May.

Overholt, William H., 1994. *The rise of China: How economic reform is creating a new superpower*. WW Norton and Company.

Pargeter, Alison. 2017. *After the fall: Views from the ground of international military intervention in post-Gadhafi Libya*. London: Remote Control Project.
https://www.oxfordresearchgroup.org.uk/after-the-fall-views-from-the-ground-of-international-military-intervention-in-post-gadhafi-libya

Rachman, Gideon. 2019. 'Trump era puts great power rivalry at centre of US foreign policy.' *Financial Times*. 21 January.

Renz, Bettina. 2016. 'Russia and "hybrid warfare."' *Contemporary Politics*, 22(3): 283–300.

———. 2018. *Russia's Military Revival*. Polity.

Saeed, Luqman. et al. 2019. *Drone Strikes and Suicide Attacks in Pakistan: An Analysis.* Action Against Armed Violence.

Sander, Alison, and Meldon Wolfgang. '2014. BCG Perspectives: The Rise of Robotics.' Boston Consulting Group.

Scharre, Paul. 2014. *Robotics on the Battlefield Part II: The Coming Swarm.* CNAS. https://www.cnas.org/publications/reports/robotics-on-the-battlefield-part-ii-the-coming-swarm

Scharre, Paul. 2019. 'The Real Dangers of an AI Arms Race.' *Foreign Affairs*, 9(3). May/June.

Schwarz, Elke. 2018. 'The (Im)possibility of Meaningful Human Control for Lethal Autonomous Weapon Systems, Humanitarian Law and Policy.' ICRC. 29 August. https://blogs.icrc.org/law-and-policy/2018/08/29/im-possibility-meaningful-human-control-lethal-autonomous-weapon-systems/

Sharkey, Noel. 2017. 'Why Robots Should Not be Delegated with the Decision to Kill.' *Connection Science*, 29(2): 177–186.

Splidsboel Hansen, Flemming. 2017. *Russian hybrid warfare: a study of disinformation* (No. 2017: 06). DIIS Report.

Stengel, Richard. 2014. 'Russia today's disinformation campaign.' *Dipnote: US Department of State Official Blog.* US Department of State.

Tamm, Hemming. 2014. *The dynamics of transnational alliances in Africa, 1990–2010: governments, rebel groups, and power politics*, PhD thesis. Oxford: University of Oxford.

Tapscott, Rebecca. 2019. 'Conceptualizing militias in Africa.' In *Oxford Research Encyclopaedia of Politics*. Oxford.

The White House. 2015. National Security Strategy of the United States of America, 2015. https://obamawhitehouse.archives.gov/sites/default/files/docs/2015_national_security_strategy_2.pdf

The White House. 2017. *National Security Strategy of the United States of America, 2017.* https://www.whitehouse.gov/wp-content/uploads/2017/12/NSS-Final-12-18-2017-0905.pdf

Thomas, Timothy. 2014. 'Russia's Information Warfare Strategy: Can the Nation Cope in Future Conflicts?' *The Journal of Slavic Military Studies*, 27(1): 101–130.

Trenin, Dmitri. 2016. 'The revival of the Russian military: How Moscow reloaded.' *Foreign Affairs*, 95(3): 23–29.

Tubiana, Jérôme, and Emily Walmsley. 2008. 'The Chad-Sudan proxy war and the "Darfurization" of Chad: myths and reality.' Working Paper. Geneva: Small Arms Survey.

United States Department of Defence. 2018. *2018 National Defense Strategy*.

Watling, Jack. and Shabibi, Namir. 2018. *Defining Remote Warfare: British Training and Assistance Programmes in Yemen, 2004–2015.* Oxford Research Group.

Watson, Abigail, and Megan Karlshøj-Pedersen. 2019. *Fusion Doctrine in Five Steps: Lessons Learned from Remote Warfare in Africa.* Oxford Research Group. https://www.oxfordresearchgroup.org.uk/fusion-doctrine-in-five-steps-lessons-learned-from-remote-warfare-in-africa

Watson, Abigail. 2020a. 'Planning for the World After COVID-19: Assessing the Domestic and International Drivers of Conflict.' Oxford Research Group. 23 April. https://www.oxfordresearchgroup.org.uk/planning-for-the-world-after-covid-19-assessing-the-domestic-and-international-drivers-of-conflict

————.. 2020b. 'Questions for the Integrated Review #2: How to Engage: Deep and Narrow or Wide and Shallow?' Oxford Research Group. 14 July. https://www.oxfordresearchgroup.org.uk/questions-for-the-integrated-review-2-how-to-engage-deep-and-narrow-or-wide-and-shallow

Watts, Tom, and Biegon, Rubrick. 2018. 'Conceptualising Remote Warfare: Past, Present and Future.' Oxford Research Group. 23 May.

World Health Organisation. 2020. 'WHO Coronavirus Disease (COVID-19) Dashboard.' https://www.worldometers.info/coronavirus/

Win, Ju, and Steven Lee Myers. 2020. 'Battle in the Himalayas.' *New York Times.* 18 June. https://www.nytimes.com/interactive/2020/07/18/world/asia/china-india-border-conflict.html

Note on Indexing

Our publications do not feature indexes. If you are reading this book in paperback and want to find a particular word or phrase you can do so by downloading a free PDF version of this book from the E-International Relations website.

View the e-book in any standard PDF reader such as Adobe Acrobat Reader (pc) or Preview (mac) and enter your search terms in the search box. You can then navigate through the search results and find what you are looking for. In practice, this method can prove much more targeted and effective than consulting an index.

If you are using apps (or devices) to read our e-books, you should also find word search functionality in those.

You can find all of our e-books at: http://www.e-ir.info/publications